T0128029

Understanding the Christianity–Evolution Relationship

The relationship between science and religion is a topic that runs rife with misconceptions, misunderstandings and debates. Are science and religion always in conflict? Is Darwin's theory of evolution through natural selection atheistic? How does history shape current debates around science and religion? This book explores these questions in a neutral and balanced way, focusing on the Christianity–evolution relationship. It shows that two paradigms – the world as an organism and the world as a machine – have critically informed and guided the discussions. The author uses his deep understanding of the history and philosophy of science, particularly Darwinian evolutionary theory and its controversies through the past 150 years, to bring fresh ideas to the debate and to wider discussions such as environmental issues and hate. *Understanding the Christianity–Evolution Relationship* provides a lively and informative analysis and lays out multiple views so that readers can make their own judgements to increase their understanding.

Michael Ruse is Professor Emeritus at the University of Guelph, Ontario and recently retired as the Lucyle T. Werkmeister Professor of Philosophy at Florida State University, Florida. His research interests focus on Darwinian evolutionary theory. He is a Guggenheim Fellow, a Gifford Lecturer, a Fellow of the Royal Society of Canada and the recipient of four honorary degrees. He is the author/editor of over 60 books, including *The Cambridge Encyclopedia of Darwin and Evolutionary Thought* (Cambridge University Press, 2013), *The Cambridge History of Atheism* (Cambridge University Press, 2021) and *Understanding Natural Selection* (Cambridge University Press, 2022).

The **Understanding Life** series is for anyone wanting an engaging and concise way into a key biological topic. Offering a multidisciplinary perspective, these accessible guides address common misconceptions and misunderstandings in a thoughtful way to help stimulate debate and encourage a more in-depth understanding. Written by leading thinkers in each field, these books are for anyone wanting an expert overview that will enable clearer thinking on each topic.

Series Editor: Kostas Kampourakis http://kampourakis.com

Published titles:

Understanding Evolution	Kostas Kampourakis	9781108746083
Understanding Coronavirus	Raul Rabadan	9781108826716
Understanding Development	Alessandro Minelli	9781108799232
Understanding Evo-Devo	Wallace Arthur	9781108819466
Understanding Genes	Kostas Kampourakis	9781108812825
Understanding DNA Ancestry	Sheldon Krimsky	9781108816038
Understanding Intelligence	Ken Richardson	9781108940368
Understanding Metaphors in the Life Sciences	Andrew S. Reynolds	9781108940498
Understanding Cancer	Robin Hesketh	9781009005999
Understanding How Science Explains the World	Kevin McCain	9781108995504
Understanding Race	Rob DeSalle and Ian Tattersall	9781009055581
Understanding Human Evolution	Ian Tattersall	9781009101998
Understanding Human Metabolism	Keith N. Frayn	9781009108522
Understanding Fertility	Gab Kovacs	9781009054164
Understanding Forensic DNA	Suzanne Bell and John M. Butler	9781009044011
Understanding Natural Selection	Michael Ruse	9781009088329
Understanding Life in the Universe	Wallace Arthur	9781009207324
Understanding Species	John S. Wilkins	9781108987196
Understanding the Christianity–Evolution Relationship	Michael Ruse	9781009277280

Forthcoming:

Understanding
the Christianity–Evolution
Relationship

MICHAEL RUSE
University of Guelph
Florida State University

CAMBRIDGE
UNIVERSITY PRESS

Shaftesbury Road, Cambridge CB2 8EA, United Kingdom

One Liberty Plaza, 20th Floor, New York, NY 10006, USA

477 Williamstown Road, Port Melbourne, VIC 3207, Australia

314–321, 3rd Floor, Plot 3, Splendor Forum, Jasola District Centre,
New Delhi – 110025, India

103 Penang Road, #05–06/07, Visioncrest Commercial, Singapore 238467

Cambridge University Press is part of Cambridge University Press & Assessment,
a department of the University of Cambridge.

We share the University's mission to contribute to society through the pursuit of
education, learning and research at the highest international levels of excellence.

www.cambridge.org
Information on this title: www.cambridge.org/9781009277280

DOI: 10.1017/9781009277334

First published 2023

A catalog record for this publication is available from the British Library.

A Cataloging-in-Publication data record for this book is available from the Library of
Congress.

ISBN 978-1-009-27728-0 Paperback

Cambridge University Press & Assessment has no responsibility for the persistence
or accuracy of URLs for external or third-party internet websites referred to in this
publication and does not guarantee that any content on such websites is, or will
remain, accurate or appropriate.

"The indefatigable Michael Ruse has produced a fascinating and most distinctive book in his *Understanding the Christianity–Evolution Relationship*. Eschewing a conventional approach to exploring this aspect of the science and religion question, Ruse uses his exceptional knowledge of the history and philosophy of biology to look at a very wide range of aspects of the Christianity–evolution relationship. These he illuminates with his inimitable turns of phrase and frequent deep insights."

Revd Professor Michael J. Reiss, University College London and the International Society for Science and Religion

"As a prolific and insightful commentator on Darwin and all things Darwinian, Michael Ruse has few, if any, equals. Devotee of modern Darwinian science, but no hater of Christianity, he offers refreshing balance by showing how both Christianity and science have been influenced, though differently, by the same rival paradigms of mechanism and organicism. Readers should not expect a deep theological treatise, but they will find a lively introduction to discourse about science and religion, written with striking informality and providing plenty of stimulus to polish their own thinking."

John Hedley Brooke, University of Oxford

"The compatibility, or not, of science and religion (specifically Christianity) is a centuries-old issue, which intensified in 1859 with Darwin's extension of the mechanistic explanation of the structure and behavior of the world around us to living things, including humans. Ruse offers, not an answer, as such, to this debate, but a skillful examination of the intellectual chess-match: moves and counter-moves. The template of his narrative centres on the mechanistic and organismic views of nature. This thread is brilliantly embellished with clear expositions of all the perspectives advanced over the last two or so centuries."

R. Paul Thompson, Professor Emeritus, University of Toronto

"In this little book, Michael Ruse reaps a huge harvest from decades of writing on the relationship between evolutionary science and Christianity. Displaying an admirable breadth of learning in both Christian theology and evolutionary biology, Ruse shows that in their best formulations, neither human enterprise needs to wage war against the other. Ongoing hostilities originate either from bad articulations of Christianity (Creationism), bellicose banishment of final causes from science (New Atheism), or (as often) both together. In thought-provoking fashion, Ruse also diffuses the differences by relativizing them as boiling up in part from conflicting root metaphors, or paradigms – mechanistic vs. organistic pictures of the world. … Above all, the book should make us consider, at least, that contrary to common opinion, an intellectually honorable peace between Darwinism and Christianity is not only possible but is advantageous to science, to Christianity, and to society as a whole."

John R. Schneider, Professor Emeritus, Theology, Calvin University

To the memory of David Murray, University of Guelph, Canada
And for Sam Huckaba, Florida State University, USA

Contents

Foreword

There are so many books about science and religion. Yet another one? OK, you may think, this one is not about science and religion in general, but about the life sciences and Christianity in particular. But so what? Rather than being about Galileo and the Inquisition, will it be about Darwin's "Bulldog" Thomas Henry Huxley defeating Bishop Samuel Wilberforce (a.k.a. "Soapy Sam") in Oxford, and the like? Well, the answer is emphatically no. The present book is like no other book on science and religion that you have ever read. Based on his 55 years of scholarship and teaching, Michael Ruse has produced a splendid, and thoughtful, account of the deeper differences and similarities between the ways that the life sciences and Christianity approach nature and humanity. Drawing on the distinction between organicism and mechanism, Ruse shows how we can find influences of both on both the life sciences and Christianity. In the end there is no right versus wrong, no bad versus good, but a complicated relationship among two human endeavors to understand nature and our place in it. This is why one can accept both, only one, or neither of them. Read this book, and your understanding of both science and religion will never be the same again.

Kostas Kampourakis, Series Editor

Preface

Evolution and Christianity? Start with evolution, or more broadly with science itself, the effort to describe and understand the natural world, understanding this to mean the mental as well as the physical. Already, constraints (or guidelines) start to appear. In the words of the eminent historian Lynn White Junior: "One thing is so certain that it seems stupid to verbalize it: both modern technology and modern science are distinctively Occidental. Our technology has absorbed elements from all over the world, notably from China; yet everywhere today, whether in Japan or in Nigeria, successful technology is Western." He does not deny the historical importance of cultures with religions other than Christianity, the dominant Western religion – Islam, in the Middle Ages, stands out. But even here, the modern debt is to such elements as these, the legacy of great thinkers, as taken into our culture: "Indeed, not a few works of such geniuses seem to have vanished in the original Arabic and to survive only in medieval Latin translations that helped to lay the foundations for later Western developments," leading to the conclusion: "Today, around the globe, all significant science is Western in style and method, whatever the pigmentation or language of the scientists."

What of the life sciences, or, as they are perhaps better known, "biology" or the "biological sciences"? This is the area of science that deals with living things, past and present: animals and plants, most obviously to us humans, but also a myriad of micro-organisms. It ranges across anatomy, the shape of things; physiology, the working of things; embryology, the growth of things; ecology, things in their environments; and most obviously, evolution, the history of things. Is each area equally important? Well, yes, in a sense it is.

But this little book is about the life sciences and Christianity, the dominant religion in that part of the world from which modern science emerged. This shapes our judgment of "importance." Christianity is a historical religion. It has a story of origins involving God and human relationships to Him, of our present obligations, and something about how it is all going to end in the future. It requires no great talent to see that the area of biology that will interact most directly with religion is the area of biology that deals with history: evolution. In an important way, this is a plus rather than a minus. The eminent biologist Theodosius Dobzhansky used to say: "Nothing in biology makes sense except in the light of evolution."

Hence, this book will focus primarily on evolution and Christianity. Primarily, but not exclusively. Throughout the book, putting things in context, there will be discussion of evolution and its relationships to other areas of biology – molecular biology and ecology, for instance. I am a philosopher and historian of science, not a theologian. As you might expect, my interests and expertise lie more with the former than the latter. Showing this immediately, I shall start right off in the first chapter discussing some philosophical aspects of science, setting them in historical context. Questions of religion will be picked up in the second chapter and continue right through to the final chapter, the sixth. Concluding, as is the custom in the series in which this book appears, there is a list of "common misunderstandings." Although I make no explicit reference to these in the book's general discussion – too many suggestions and directions make for stilted reading – I have the misunderstandings very much in mind. For this reason, I suggest that readers start this book by looking at the misunderstandings, and then you yourselves can figure out my answers to the questions they pose. I doubt you will agree with all my answers. As a teacher for over 50 years, I shall feel my task is well done if I stimulate you to work out answers that you think correct – and perhaps ask questions that I overlooked. If you agree with everything I say, I shall not think you are taking me seriously.

Acknowledgements

Three people have been very influential on me, guiding and informing my thinking. First, the late Edward O. Wilson, the pre-eminent evolutionist of his day, a pioneer in the field of sociobiology, the study of social behavior from an evolutionary perspective. For over 40 years, he was a mentor and friend. Second and third, Robert J. Richards at the University of Chicago and Joseph Cain at University College London. Friends also, the former has helped me to understand the history of evolutionary theory in the nineteenth century, and the latter has helped me to understand the history of evolutionary theory in the twentieth century. The Institute on Religion in an Age of Science has an annual, week-long conference on Star Island off the coast of New Hampshire. The members, mainly Unitarian-Universalists or members of the United Church of Christ, are as modest as the title of their organization is pretentious. I have been part of the group for 40 years now, welcomed in although my non-belief is known to all. I cannot thank them enough for their friendship and friendly, if critical, support of my journey through this vale of soul making.

I am deeply grateful to my series editor, Kostas Kampourakis, as well as to the team at Cambridge University Press. Special mention must be made of my press editor Olivia Boult and my copyeditor Lindsay Nightingale. Working with such people as these is a joy and a privilege. As always, my chief debt is to my wife Lizzie, ever there to give me love and encouragement, which does not preclude warning me when I am going over the top. My dedication is to two of the deans under whom I have served. The first is to the late David Murray at the University of Guelph and the second to Sam Huckaba at Florida State

University. I choose them both as friends and leaders in themselves, but also as representatives of my many colleagues over 55 years of teaching who made it all so very much worthwhile. This book is a small token of my thanks.

Finally, no acknowledgements by me would be complete without mention of my cairn terriers, Scruffy McGruff and Duncan Donut. They are ever ready to tell me to drop tools and go for a walk in the park, which, on reflection, is not a bad idea right now before I start into the main text of this book.

1 Rival Paradigms

"Just the facts, ma'am. Just the facts!" This famous directive by Sergeant Joe Friday – apparently never actually made in this form – is from the television series *Dragnet*. Unfortunately, while this may be adequate for detecting and solving crime, not so elsewhere. The idea that science is simply a matter of recording empirical experience is hopelessly inadequate and misleading. Science is about empirical experience, but it is about such experience as encountered and interpreted – and with effort and good fortune – as explained by us. To this end, we view the world, external and internal, through the lenses, as it were, of modes of understanding. Above all, metaphorical modes of understanding. In scientific thinking, there have been two major metaphors: what linguists call "root metaphors," what – borrowing and somewhat extending the ideas of Thomas Kuhn – philosophers call "paradigms." Two world interpretive visions. There is the root metaphor or paradigm of the world and its parts as organisms. The organic paradigm. *Organicism*. And there is the root metaphor or paradigm of the world and its parts as machines. The machine paradigm. *Mechanism*. These metaphors or paradigms and their differences will structure the discussion of this book. Let's get straight to work, looking at the metaphors in their historical contexts.

Plato and Aristotle

The organic metaphor was the dominant vision for the Ancient Greeks. No surprise, really. It is nigh impossible to give accurate population sizes, but around 400 BCE, the time of the great philosophers, there were about two million people in Greece proper – considerably more if you count all the Greek-settled areas (like Sicily). The population of its biggest city, Athens, was about 150,000, taking in slaves and foreigners and the like. Including suburbs, twice that size. Even by the most generous estimate, the important point is that most people lived in rural areas, close to the land and the heavens, particularly the night sky in a land with no technically advanced lighting. It was natural to think in organic terms. Spring, birth, and the early years; summer, growth to full maturity; autumn, appreciating one's achievements, but slowing down; winter, death, but with the prospect of renewal and spring again, generation after generation. And the parts and processes of the world can be given an organic interpretation. Water, the life blood – rain, fertilization, rivers carrying things away, lakes, seas, oceans. All can be understood in organic terms. One must think in terms of wholes – what the soldier, statesman, philosopher, the South African Jan Smuts, at the beginning of the twentieth century was to call "holism."

Plato presented this vision somewhat more formally in his dialogue *Timaeus* – "more formally" in the sense that Plato presented his thinking against the background of his metaphysical Theory of Forms. Especially in the *Republic*, Plato argued that this world of ours is one of change, transient, and a kind of state of being between nothingness below and mathematics and the Forms above. Forms have many roles – too many at times – but they are standards and also function as universals. "Dobbin" is the individual; "horse" is the Form. These are the truly real and they exist in a kind of world of rationality, and, as the truths of mathematics, eternal, unchanging. The Forms are ordered, and at the top, giving life to all the others, is the Form of the Good. Much influenced by Pythagorean thinking, Plato likened the Form of the Good in the world of rationality to the sun in our world of change. Just as the world thrives and has its ultimate being because of the sun, so likewise the Forms have their ultimate being because of the Form of the Good (*Plato: Complete Works*).

The *Timaeus* accepts this thinking as background. The world of the Forms is unchanging and good. Our world, the world of becoming, owes its existence to the world of the Forms. The Creator made the world an organism, so that it could be as good, as perfect, as possible. It is valuable:

> God desired that all things should be good and nothing bad, so far as this
> was attainable . . . . For which reason, when he was framing the universe,
> he put intelligence in soul, and soul in body, that he might be the creator

of a work which was by nature fairest and best. Wherefore, using the language of probability, we may say that the world became a living creature truly endowed with soul and intelligence by the providence of God. (*Timaeus* 30b, in *Plato: Complete Works*)

What is the nature and status of this Creator? A kind of principle of ordering, identical with or perhaps emanating from the Good, in the *Timaeus* called the "Demiurge." From the Good come the other Forms, hence it is the Forms in general on which our world is patterned. "Well, if this world of ours is beautiful and its craftsman good, then clearly he looked at the eternal model." The oak tree is good because it is modeled on – what Plato in the *Republic* says "partici-pates" in – the Form of the Oak. But why should we think or judge this way? What is the fairest and best, the beautiful? In the *Phaedo*, Plato makes it clear that he is thinking in terms of ends, of what today is known as "teleology." You cannot understand just in terms of things happening. You must ask about results.

If mind is the disposer, mind will dispose all for the best, and put each particular in the best place; and I argued that if any one desired to find out the cause of the generation or destruction or existence of anything, he must find out what state of being or doing or suffering was best for that thing, and therefore a man had only to consider the best for himself and others, and then he would also know the worst, since the same science comprehended both. (*Phaedo*, 97 c–d)

The Good wanted to make the world as good as it could be. To do so, it had to make the world into an organism. But why would this be the best? Because this is the most desirable.

Turn now to Plato's student, follower, and critic, Aristotle. Like Plato he saw a being, or rather a Being, as the secret behind, the cause of, the way the world works. Like Plato, he saw (as a consequence) the need and possibility of explaining things in terms of their ends – teleology. "Nature never makes anything without purpose." But from there, the differences could not be starker. Aristotle's God or Creative force, known as the "Unmoved Mover," is the cause of everything. It is the ultimate Being, that which is cause of itself and infinitely good. "The first mover, then, of necessity exists; and in so far as it is necessary, it is good, and in this sense a first principle" (*Metaphysics* 1072b10–11). It is that which motivates everything.

> There is, then, something which is always moved with an unceasing motion, which is motion in a circle; and this is plain not in theory only, but in fact. Therefore, the first heavens must be eternal. There is therefore also something which moves them. And since that which is moved and moves is intermediate, there is a mover who moves without being moved, being eternal, substance, and actuality. (*Metaphysics* 1072a22–6)

The rest of existence is directed toward the Unmoved Mover, wanting in some sense to get close to it and share the perfection. Reproduction has a key role here. Organisms do not become eternal. However, through reproduction, they get as close to the eternal as possible, and that in itself is a good.

> The acts in which [the soul] manifests itself are reproduction and the use of food, because for any living thing that has reached its normal development … the most natural act is the production of another like itself, an animal producing an animal, a plant a plant, in order that, as far as nature allows, it may partake in the eternal and divine. That is the goal to which all things strive, that for the sake of which they do whatsoever their nature renders possible. (*De Anima* 415a25–415b1)

Not only in the nature of the ultimate Being but in the way the system works, Aristotle differs significantly from Plato. They both think in terms of ends, but whereas for Plato the ends come from the Designer – external teleology – for Aristotle the ends come from within, they are produced by the way that things are – internal teleology.

Famously, in his *Metaphysics*, Aristotle distinguished four kinds of cause. Consider making a statue, for example a British foot soldier – a "Tommy" – from the First World War. You have the *efficient* cause, the modeler or sculptor who actually made the statue. You have the *material* cause, the substance from which it is made (bronze or marble or what?). You have the *formal* cause, somewhat akin to a Platonic Idea (without committing oneself to the reality of such an Idea). You would not have the soldier wearing a *Pickelhaube* (German helmet with a spike). And last, but far from least, you have the *final* cause, the teleological element giving the reason for the statue. Why is the statue being made now? So that future generations can remember and give thanks for the sacrifices of him and his comrades. Note something distinctive about final causes as opposed to the other causes. An efficient cause is happening now to make a statue now for remembrance later. Even if no one ever saw the statue, it would still have the efficient cause of the modeler or the sculptor. In the case of a final cause, however, the reference is to the future, and there is always the chance that that future may never occur. An accident on the way to the memorial site means the statue is destroyed and never brings on memories. This is known as the "missing goal object." In the case of external teleology, it is the idea that counts, and this in its way is an efficient cause. It refers to the future – let's make a statue to honor our troops – but it is a reference, not the actual future. In the case of internal teleology, no such easy escape. You just have to say that nature is inherently teleological, even if things don't work out as hoped and expected.

One final question of both Plato and Aristotle. What about our own species? What about human beings? Organisms grow, from oak to acorn, from tadpole to frog. There is direction and usually, if not always, it is thought to be a progressive direction, from lesser to greater, from little worth to great worth, from "monad to man." One expects to find – one would be flabbergasted not to find – that our two philosophers agreed entirely with this summation. As so it proves.

> God gave the sovereign part of the human soul to be the divinity of each one, being that part which, as we say, dwells at the top of the body, inasmuch as we are a plant not of an earthly but of a heavenly growth, raises us from earth to our kindred who are in heaven. And in this we say truly; for the divine power suspended the head and root of us from that place where the generation of the soul first began, and thus made the whole body upright. (*Timaeus* 90b)

Not much ambiguity there. Nor is there in Aristotle. We may infer "that, after the birth of animals, plants exist for their sake, and that the other animals exist for the sake of man . . .. Now if nature makes nothing incomplete, and nothing in vain, the inference must be that she has made all animals for the sake of man" (*Metaphysics*, 1256b15–22). Likewise, explaining why humans alone are bipedal: "of all living beings with which we are acquainted man alone partakes of the divine, or at any rate partakes of it in a fuller measure than the rest." Hence, "in him alone do the natural parts hold the natural position; his upper part being turned towards that which is upper in the universe. For, of all animals, man alone stands erect" (656a17–13).

The Atomists

Did no one in the Ancient World want to challenge this teleology-impregnated view of the universe? As it happens, from the beginning – before Plato and Aristotle – there was a school of thought that wanted nothing to do with final causes. The pre-Socratic atomists – Leucippus, Democritus, and a little later Epicurus – believed that the world is made up of minute physical particles, buzzing around in the void, in empty space. Efficient causation explains all. Final cause thinking doesn't have a dog in the fight. The best account of this philosophy came some centuries later from the pen of the Roman poet Lucretius. Laying things out in *On the Nature of Things* (*De Rerum Natura*), he focused on development, not just of individual organisms but of whole groups or species. Everything came about through blind chance, with no purpose or end thinking needed (quoted by Sedley in *Creationism*, 150–3):

> At that time the earth tried to create many monsters
> with weird appearance and anatomy –
> androgynous, of neither one sex nor the other but somewhere in
>   between;
> some footless, or handless;
> many even without mouths, or without eyes and blind;
> some with their limbs stuck together all along their body,
> and thus disabled from doing harm or obtaining anything they
>   needed.
> These and other monsters the earth created.
> But to no avail, since nature prohibited their development.
> They were unable to reach the goal of their maturity,
> to find sustenance or to copulate.
>
> (*De rerum natura* V 837–48)

A hotchpotch individual thus formed, three legs, one attached to the back between the shoulders, no mouth or eyes but with six pairs of ears, was not going to last long. However, given time enough, even the improbable becomes actual.

> First, the fierce and savage lion species
> has been protected by its courage, foxes by cunning, deer by speed of
>   flight.

> But as for the light-sleeping minds of dogs, with their faithful heart,
> and every kind born of the seed of beasts of burden,
> and along with them the wool-bearing flocks and the horned tribes,
> they have all been entrusted to the care of the human race . . .   (V 862–7)

Only efficient causes here. No final causes. Eyes were not made for seeing or legs for walking. First came the eyes and legs, and then they were put to use. Denying this is to get things backwards:

> All other explanations of this type which they offer
> are back to front, due to distorted reasoning.
> For nothing has been engendered in our body in order that we might be
>   able
> to use it.
> It is the fact of its being engendered that creates its use.   (V 832–5)

It scarcely needs saying that, ingenious though this may be, it hardly convinced anyone. Even given nigh infinite time, functioning eyes and mouths, arms and legs are not going to appear on the scene. Elephants don't fly; arms and legs do not appear by chance. An adequate approach, including one like the atomists', that wants nothing to do with Creators or Unmoved Movers or the like, must still explain final cause – not downplay or ignore it.

The Christians

With the arrival of Christianity, which sees everything in terms of ends, there was even less reason for atomism to make headway. The organicist paradigm is tailor-made for Christianity. It stresses the unity of all existence, central to the Christian vision, where all comes from and ever depends on God. "Great is our Lord and abundant in strength; His understanding is infinite" (Psalm 147:5). The world is of great value and worth. "And God saw every thing that he had made, and, behold, it was very good" (Genesis 1, 31). And, most importantly, all is temporal and there is an advance through time: acorn to oak; monad to man. "So God created man in his own image, in the image of God created he him; male and female created he them" (Genesis 1, 27) (Fig. 1.1).

Figure 1.1   God creating man in His own image. (Michelangelo)

Note that God created. Hence, things do not have value in themselves. It comes from God. There is value, but it is imputed not discovered. To quote Calvin:

> And concerning inanimate objects, we ought to hold that, although each one has by nature been endowed with its own property, yet it does not exercise its own power except in so far as it is directed by God's ever-present hand. These are, thus, nothing but instruments to which God continually imparts as much effectiveness as he wills, and according to his own purpose bends and turns them to either one action or another. (*Institutes*, 1, 16, 2)

Calvin was deeply influenced by the fourth-century Roman theologian St Augustine of Hippo, and what he wrote is equally precisely the position of neo-Augustinians today. "The earth is very good. Neither demonic nor divine, neither meaningless nor sufficient unto itself, it receives its meaning and value from God," according to the Evangelical Lutheran Church in America (*This Sacred Earth*, 245).

Augustine, the very greatest of the early Christian theologians/philosophers, was an ardent Platonist, albeit at second-hand through the Hellenistic philosopher, Plotinus. In his *Confessions*, Augustine's characterization of God could have come straight out of the *Republic*. Necessary: "For God's will is not a creature but is prior to the created order, since nothing would be created unless the Creator's will preceded it. Therefore God's will belongs to his very substance."

Existing outside space: "no physical entity existed before heaven and earth." Outside time: "Your 'years' neither come nor go. Our years come and go so that all may come in succession. All your 'years' exist in simultaneity, because they do not change; those going away are not thrust out by those coming in ... Your Today is eternity."

Faith is always going to be first for Christians. Yet it was hardly going to be the case that someone of Augustine's incredible philosophical ability was going to turn his back on evidence and reason – what is known as "natural theology" as opposed to "revealed theology" or "religion" – and no more does he. He picks up what is known as the argument from design. "Even leaving aside the voices of the prophets, the world itself, by the perfect order of its changes and motions, by the great beauty of all things visible, claims by a kind of silent testimony of its own both that it has been created, and also that it could not have been made other than by a God ineffable and invisible in greatness, and ineffable and invisible in beauty" (*Confessions*, 53). Ours is a world of great value, created intentionally by a loving God.

There is one potentially awkward point that needs attention. Religions tend to have their sacred books, the truths of which are taken as absolute. In the case of Christianity, it is the Holy Bible – Old and New Testaments. Yet within its pages, particularly in the early chapters of Genesis, there are claims that must be taken on faith, but sit uncomfortably with reason. Even if reason does not have the all-conquering power it might have been thought to have, it is still important and needs attention. How do we deal with biblical claims, especially those claims about the biblical order of creation, that seem completely impossible, from the viewpoint of reason? Genesis tells us that light and dark were created on the First Day, but that we had to wait for the Fourth Day for the sun to make an appearance. Impossible! Augustine's solution was very modern-sounding, or perhaps more generously we should say that our solution is very Augustinian-sounding. He argued that the Bible is true, through and through. But sometimes it is necessary to interpret it allegorically. Why? Well, for a start, the Ancient Jews were on the whole illiterate. They were not sophisticated thinkers like fourth-century CE Romans. Too literal, and they wouldn't understand a word that was going on. So, God tempered the wind to the shorn lamb – or Israelite. God created, probably all at one time, and then explicated in a way that we can catch the important truths.

St Augustine laid the foundation. Others built on this, especially in the realm of natural theology – most famously, St Thomas Aquinas. Like Augustine, accepting that God is the creative cause of all that exists, in countering the classic undergraduate counter – "What caused God?" – Aquinas argues that God has no need of a cause. He is outside time and space. He exists necessarily. It is part of His being that He cannot not exist. This is aseity: "it affirms that God is completely self-sufficient, having within Godself the sufficient reason for God's own existence" (*New Catholic Encyclopedia*). This is not intended to be something new or radical. Far from it. It endorses the Augustinian position that God is not just eternal, but unchanging. Where Aquinas is distinctive is that he is much influenced by Aristotle, whose works were only now being translated from Greek to Latin. His thinking tended more toward internal teleology than external teleology. God works more as a principle of ordering than as an intervening hands-on designer. Either way, as we move out of the medieval era, organicism rules okay.

The Machine Metaphor

According to historian of science Eduard Jan Dijksterhuis:

> At all times there used to be a strong tendency among physicists, particularly in England, to form as concrete a picture as possible of the physical reality behind the phenomena, the not directly perceptible cause of that which can be perceived by the senses; they were always looking for hidden mechanisms, and in so doing supposed, without being concerned about this assumption, that these would be essentially the same kind as the simple instruments which men had used from time immemorial to relieve their work, so that a skillful mechanical engineer would be able to imitate the real course of the events taking place in the microcosm in a mechanical model on a larger scale. (*Mechanization*)

A new root metaphor. The world as a machine. We are coming now into the sixteenth and seventeenth centuries. Why would the organism metaphor be falling out of favor, and why then would the machine metaphor be taking over? Because, on the one hand, society – European society – became far less rural and far more urban. The immediate appeal of organicism diminished. On the other hand, more positively, machines did start to come into their own!

Their natures and virtues were becoming apparent. Above all, there was the watch or clock. They worked on and on, governed only by unbroken laws. In his *A Free Enquiry into the Vulgarly Received Notion of Nature* Robert Boyle, seventeenth-century chemist and philosopher, spelt things out: the world is

> like a rare clock, such as may be that at Strasbourg, where all things are so skillfully contrived that the engine being once set a-moving, all things proceed according to the artificer's first design, and the motions of the little statues that as such hours perform these or those motions do not require (like those of puppets) the peculiar interposing of the artificer or any intelligent agent employed by him, but perform their functions on particular occasions by virtue of the general and primitive contrivance of the whole engine. (Fig. 1.2)

Figure 1.2   Strasbourg clock.

Clocks have purposes, final causes – to tell the time. Likewise for other machines. A guillotine is for chopping off heads, a pump is for getting water out of the ground. However, within the explanation – inasmuch as it is a scientific explanation – there are no purposes, no ends, no final causes. All that matters is that the clock goes round and round, without interference, governed by blind, purposeless laws. The Earth, perhaps, may have been designed by the Demiurge as an abode for human beings, but under the machine metaphor that is extraneous. It could have readily been a place of punishment for fallen angels – let them see what it is like to have to take a logic course. No final causes, and hence no values. The clock has value to us; but, in itself, it just is. Neither good nor bad. A grand new picture. The universe – the heliocentric universe of Copernicus – is easily seen as a huge clock. As the parts of the clock go through their motions, endlessly, without purpose in the system, so the parts of the universe, better known as stars and planets, go through their motions, endlessly, without purpose in the system. God or some other force like the Unmoved Mover may be responsible for it all, but that kind of discussion belongs to theology not science. As the philosopher Francis Bacon said wittily, final causes were like Vestal Virgins, beautiful but barren. God had become a "retired engineer."

But was He? Compare the solar system with the human body. Agree that both are designed and made by God for the benefit of humankind. Ask about the purpose of the moon. You can joke that it exists to light the way home for drunken philosophers; but, in fact, although there was an eighteenth-century society that did meet at full moon for purposes of getting home safely (the Lunar Society), it is really a joke. The moon within the solar system has no purpose, no function. If it were removed, things would just keep going on. Same with the planets. Now ask about the purpose of the heart. Within the system it does have a purpose. It exists for pumping blood, which exists for capturing oxygen and all that that does for us. It seems that when it comes to animate matter, organisms, final cause understanding does have a role. Which rather suggests that the machine metaphor is not all encompassing. It may work for the solar system – one thing after another – but it does not work for organisms – one thing in order for another. More accurately, we should say that the machine metaphor is not adequate for organisms. The human heart is a machine – for pumping blood – but this does not answer the question of why it should pump blood. There is a design-like odor about the animate that one does not sense about the inanimate (Fig. 1.3).

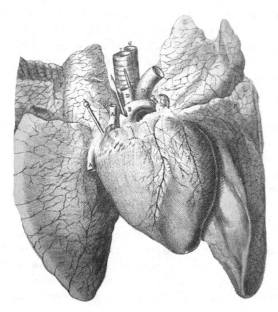

Figure 1.3 The heart, from René Descartes' *De Homine* (1662).

Robert Boyle saw this problem, and agreeing that raw atomism goes nowhere, offered a solution. Talk of mechanisms is part of science. Talk of final causes is part of theology! In his *Disquisition about the Final Causes of Natural Things*, satisfyingly making a philosophical point while putting the boot into the French, he wrote:

> For there are some things in nature so curiously contrived, and so exquis-itely fitted for certain operations and uses, that it seems little less than blindness in him, that acknowledges, with the Cartesians [followers of Descartes], a most wise Author of things, not to conclude, that, though they may have been designed for other (and perhaps higher) uses, yet they were designed for this use.

Boyle continued that the supposition that "a man's eyes were made by chance, argues, that they need have no relation to a designing agent; and the use, that a man makes of them, may be either casual too, or at least may be an effect of his knowledge, not of nature's." However, the cost of taking us away from a designing intelligence is taking us from the chance to do science – the urge to dissect and to understand how the eye "is as exquisitely fitted to be an organ of sight, as the best artificer in the world could have framed a little engine, purposely and mainly designed for the use of seeing."

Somewhat brazenly, Boyle distinguished between acknowledging the use of final causes qua science and the inference qua theology from final causes to a designing god. First: "In the bodies of animals it is oftentimes allowable for a naturalist, from the manifest and apposite uses of the parts, to collect some of the particular ends, to which nature destinated them. And in some cases we may, from the known natures, as well as from the structure, of the parts, ground probable conjectures (both affirmative and negative) about the particular offices of the parts." Then, second, the science finished, one can change tracks into theology: "It is rational, from the manifest fitness of some things to cosmical or animal ends or uses, to infer, that they were framed or ordained in reference thereunto by an intelligent and designing agent." We go from a scientific study of what Boyle called "contrivance," in the domain of science, to inferences about design – or rather Design – in the domain of theology.

Natural Theology

A compromise – mechanism is retained albeit its scope seems reduced – but one that led to a century or more of natural history, not to mention laboratory studies in morphology and embryology. It gave credence also to a vigorous strain of natural theology. The existence of God follows from the nature of organisms. From our perspective, biology and Christianity, far from being at war, are symbiotically entwined. The machine metaphor sweeps all before it in the inanimate world. It applies and explains in the animate world, but only partially. It fails to explain the design-like nature of organisms and hence – this seeming to be the only possible solution – it opens the way to belief in a god, at least in major respects akin to the God of the Christians. Unsurprisingly, from Boyle until Charles Darwin's *On the Origin of Species*, published in 1859, the argument from design flourished. Nigh every move made in the life sciences seemed to support this argument. It is little wonder then that, in 1802, without any sense of anachronism, Archdeacon William Paley published his classic exposition of the argument in his *Natural Theology*. In one of the best-known passages in philosophy – or theology if you are intent on placing the blame elsewhere – Paley invites you to compare a stone with a watch.

> In crossing a heath, suppose I pitched my foot against a *stone*, and were asked how the stone came to be there, I might possibly answer, that for any thing I knew to the contrary, it had lain there for ever; nor would it, perhaps, be very easy to show the absurdity of this answer. But suppose I had found a *watch* upon the ground, and it should be inquired how the watch happened to be in that place, I should hardly think of the answer which I had before given – that, for any thing I knew, the watch might have always been there. Yet why should not this answer serve for the watch as well as for the stone; why is it not as admissible in the second case as in the first? For this reason, and for no other, namely, that, when we come to inspect the watch we perceive – what we could not discover in the stone – that its several parts are framed and put together for a purpose, e.g. that they are so formed and adjusted as to produce motion, and that motion so regulated as to point out the hour of the day;

We are on the way to God. Paley points out that the eye is like a telescope. "Telescopes have telescope designers and makers. So, likewise, eyes have designers and makers. God!"

Not everyone, not every believer, was comfortable with this kind of argument. A few years before Paley, Immanuel Kant in his *Critique of the Power of Judgment* had argued that organisms are just machines, but that we need final-cause thinking as a heuristic guide. They help us think about organisms. They are "regulative." They are not part of reality. They are not "constitutive." They are "for guiding research into objects of this kind and thinking over their highest ground in accordance with a remote analogy with our own causality in accordance with ends." All very well. But it does mean that biology is forever condemned to be second rate. "[W]e can boldly say that it would be absurd for humans even to make such an attempt or to hope that there may yet arise a Newton who could make comprehensible even the generation of a blade of grass according to natural laws that no intention has ordered; rather, we must absolutely deny this insight to human beings."

Biology is all science, not (Boyle-like) science and religion, but the cost is that biology is second-rate science. The scientist of the organic world, faced with final causes, must necessarily live with and only with science; nevertheless, the science of the biologist can never equal the science of the physicist.

Evolution Arrives

By the end of the eighteenth century, evolutionary ideas were becoming familiar if not very enthusiastically accepted. One of the best-known who argued this way was the British physician and poet, grandfather of Charles Darwin, Erasmus Darwin. He was (as were many) much taken with the progress that was being made in Britain in the realm of industry, and this enthusiasm found its way into his poetry, as he likened advance in the human world to advance in the organic world. In his poem *The Temple of Nature*, Erasmus Darwin wrote:

> Organic Life beneath the shoreless waves
> Was born and nurs'd in Ocean's pearly caves;
> First forms minute, unseen by spheric glass,
> Move on the mud, or pierce the watery mass;
> These, as successive generations bloom,
> New powers acquire, and larger limbs assume;
> Whence countless groups of vegetation spring,
>   And breathing realms of fin, and feet, and wing.

And so down to – or, rather, up to – humans.

> Imperious man, who rules the bestial crowd,
> Of language, reason, and reflection proud,
> With brow erect who scorns this earthy sod,
> And styles himself the image of his God;
> Arose from rudiments of form and sense,
> An embryon point, or microscopic ens!

Explicitly, Darwin tied his biology to his philosophy. The idea of organic progressive evolution, as he wrote in his *Zoonomia*, "is analogous to the improving excellence observable in every part of the creation; such as the progressive increase of the wisdom and happiness of its inhabitants."

Erasmus Darwin was a little casual about the forces, the causes, that brings all of this about, but a major factor was what came to be known, after the endorsement of the French evolutionist Jean Baptiste de Lamarck (1809), as "Lamarckism." The inheritance of acquired characteristics. The blacksmith gets strong arms through working at the forge, so his son is born with such strong arms. Note that, although Erasmus Darwin was an evolutionist, he was not a strict mechanist – nor was Lamarck for that matter. They both saw an upwards progression to life, akin to the Great Chain of Being (Fig. 1.4). They saw, out there in the world, organisms getting of greater and greater value. Apes are of more value than reptiles, and humans of more value than apes. And this, in the eyes of more respectable scientists who were Christian, was where it all came unstuck. The big problem with evolution, tied as it was to progress, was the extent to which this underlying philosophy was thought incompatible with the essential underlying philosophy of Christianity – Providence. It was a mainstay of Christian thought – Protestant Christian thought particularly – that we can do nothing save for the grace of God. On our own we are helpless, as is spelt out by the popular hymn of the Congregationalist Isaac Watts.

> When I survey the wondrous cross
> On which the Prince of glory died,
> My richest gain I count but loss,
> And pour contempt on all my pride.

Progress goes directly against this, for its central theme is that we ourselves can improve things through our own intelligence and effort. No need of God.

Combine this philosophy/theology highlighting Providence with the empirical evidence – most obviously that the fossil record showed gaps between forms, quite contrary to what evolution would expect us to find – and the case was

Figure 1.4 The Great Chain of Being: God, angels, heaven, humans, beasts, plants, flame, rocks.

complete. William Whewell, general man of science and author of works on the history and philosophy of science, gave the definitive judgment: "Geology and astronomy are, of themselves, incapable of giving us any distinct and satisfactory account of the origin of the universe, or of its parts," continuing: "The mystery of creation is not within the range of her legitimate territory; she says nothing, but she points upwards" (*History* 3, 587–8). Science and religion come together harmoniously to show that evolution is not true.

This is the background to the efforts of that very ambitious young man, Charles Darwin – born on the same day as Abraham Lincoln (February 12, 1809) – who very much did want to be the Newton of the blade of grass. As we turn to him, though, it is well to remember the important point made by Thomas Kuhn in his *Structure of Scientific Revolutions*. A change in paradigms, a change in root metaphors, is rarely if ever fueled solely by the attractions of the new paradigm. Usually, if not always, the old paradigm is running into problems, internal contradictions and the like. This was very

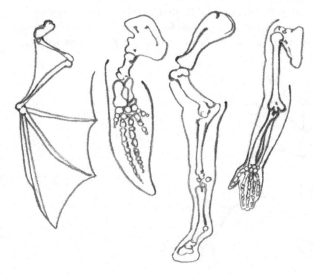

Figure 1.5   Homology.

much the case here. The problem lay in homologies: the non-functional isomorphisms between organisms of different species. Best known are the similarities between the forearm of humans, the front leg of the horse, the wing of the bat, and the flipper of the porpoise (Fig. 1.5). There seems to be no purpose here because the bodily parts all have different functions. Kuhn is right. Something was rotten in the state of Denmark, otherwise known as the argument against evolution.

Charles Darwin

Charles Darwin wanted to give an entirely mechanistic picture of the evolutionary process – unguided laws, governing a world ever in motion, leading to a "tree of life" (Fig. 1.6). He recognized that farmers and fanciers could change organisms – plumper pigs, shaggier sheep, fiercer fighting dogs, more melodious songbirds (Fig. 1.7). This was done consciously through selection of desirable qualities. How then were we to get a non-conscious, law-bound equivalent process in the natural world? Darwin started with the observation of the English clergyman, Thomas Robert Malthus, that organisms always reproduce at a faster rate than food and space can maintain. There is, therefore, an ongoing struggle for existence – more importantly, struggle for reproduction. To this, Darwin added that random variation was the norm in natural populations – he had spent

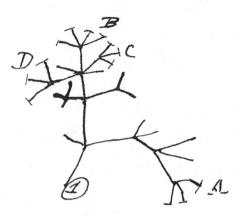

Figure 1.6   Darwin's tree of life.

eight years dissecting barnacles, so he knew whereof he spoke. Then he was ready for his main argument. "Can the principle of selection, which we have seen is so potent in the hands of man, apply in nature?" Darwin thought that it could indeed apply.

> Let it be borne in mind how infinitely complex and close-fitting are the mutual relations of all organic beings to each other and to their physical conditions of life. Can it, then, be thought improbable, seeing that variations useful to man have undoubtedly occurred, that other variations useful in some way to each being in the great and complex battle of life, should sometimes occur in the course of thousands of generations? If such do occur, can we doubt (remembering that many more individuals are born than can possibly survive) that individuals having any advantage, however slight, over others, would have the best chance of surviving and of procreating their kind? On the other hand, we may feel sure that any variation in the least degree injurious would be rigidly destroyed. This preservation of favourable variations and the rejection of injurious variations, I call Natural Selection. (*Origin*, 80–1)

The all-important point is that this was not just change, but change in the direction of adaptation.

Figure 1.7   Pig for breeding.

We see these beautiful co-adaptations most plainly in the woodpecker and missletoe; and only a little less plainly in the humblest parasite which clings to the hairs of a quadruped or feathers of a bird; in the structure of the beetle which dives through the water; in the plumed seed which is wafted by the gentlest breeze; in short, we see beautiful adaptations everywhere and in every part of the organic world. (60–1)

Natural selection. Adaptation caused by law, by non-directed law. Note that the last thing that Darwin was about was expelling teleology from biology. Going that way would lead to the implausible position of the atomists. What Darwin wanted to do was to explain it under the machine paradigm. For this reason, he comfortably and repeatedly used the term "final cause." Talking of cuckoos laying their eggs in the nests of other birds: "It is now commonly admitted that the more immediate and *final cause* of the cuckoo's instinct is, that she lays her eggs, not daily, but at intervals of two or three days; so that, if she were to make her own nest and sit on her own eggs, those first laid would have to be left for some time unincubated, or there would be eggs and young birds of different ages in the same nest" (216–17, my italics). To avoid this, the female cuckoo lays her eggs in the nests of other birds, so they can get immediate attention.

Darwin was (very satisfyingly) withering about attempts to explain away homology. "Nothing can be more hopeless than to attempt to explain this similarity of pattern in members of the same class, by utility or by the doctrine of final causes" (435). He did not deny homology. Rather, it is adaptation first, homology second. "On my theory, unity of type is explained by unity of descent" (206). Not that Darwin wanted to use this conclusion to promote atheism. Darwin at Cambridge, intending to be ordained into the Church of England, was a theist – God prepared to intervene in his Creation, as the sending of Jesus. Then, from the time of the *Beagle* voyage, Darwin became a deist – God as Creator but one who then lets everything unfold according to unbroken law. In the *Origin*, Darwin made it very clear that he saw nothing in his theory that challenged this position. Indeed, it confirmed it.

Authors of the highest eminence seem to be fully satisfied with the view that each species has been independently created. To my mind it accords better with what we know of the laws impressed on matter by the Creator, that the production and extinction of the past and present inhabitants of the world should have been due to secondary causes, like those determining the birth and death of the individual. (488–9)

Later, around 1865, Darwin became an agnostic. But like most Victorian agnostics, and Darwin was in ever-growing company, his move was theological not science-based. As he wrote in his autobiography, he had no time for Christianity because it implies that non-believers will go to hell: "this is a damnable doctrine." For Darwin, this was personal. He venerated his father, a non-believer who could have given Richard Dawkins a run for his money, as one of the finest people he knew.

After Darwin

Move on now from Darwin down to the present. The significant scientific moves were the development of an adequate theory of heredity – genetics – and its melding with Darwinian selection. This happened around 1930, thanks to the population geneticists, notably in England Ronald A. Fisher and J. B. S. Haldane, and in America Sewall Wright. A few years later, the naturalists and experimentalists got to work, and empirical flesh was put on the mathematical skeleton. Thus was born "neo-Darwinism," as it was called in Britain, and the "Synthetic Theory," as it was called in America. Notable works were, in Britain, *Evolution: The Modern Synthesis* by Julian Huxley (the grandson of Darwin's great supporter Thomas Henry Huxley, and older brother of Aldous Huxley), and, in America, *Genetics and the Origin of Species*, by the Russian-born Theodosius Dobzhansky. Things were now in place, although obviously there were ongoing changes and developments – above all, the coming of molecular biology.

The double helix, the discovery of the structure of the DNA molecule by James Watson and Francis Crick in 1953, was a triumph of mechanism, as it was shown that the molecule works on exactly the same principles of already-developed machines. Consider the Enigma Machine, used by the Germans in the Second World War to code their messages (Fig. 1.8). It functioned through a series of rotors that took in the information and

(a)

(b)

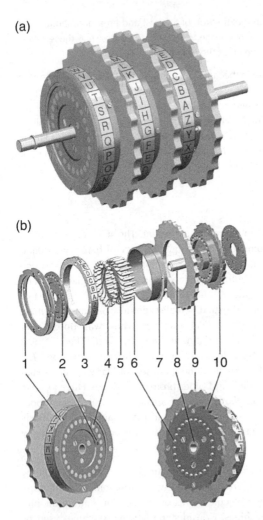

Figure 1.8   a, Enigma machine. **b**, Enigma machine parts. 1, Notched ring; 2, marking dot for A contact; 3, alphabet tire; 4, plate contacts; 5, wire connections; 6, pin contacts; 7, spring-loaded ring adjusting lever; 8, hub; 9, finger wheel; 10, ratchet wheel.

scrambled it around, so that only those with the right codes could unpack it and read what was being sent. Invented in the 1920s, in the years following, first the Poles and then the British worked to find how the machine worked and how it could be deconstructed, as it were. Vital was the recovery of an actual Enigma Machine, which could then be taken apart and examined for its functions. In other words, we have a reductive process, as it is shown how the whole machine works in terms of its constituent parts.

Turn now to the unraveling of the DNA molecule (Fig. 1.9). Watson and Crick worked in exactly the same way as did those who worked on the Enigma Machine. They took the molecule apart and then explained how it worked in terms of its constituent parts. Reduction! A DNA molecule is essentially a chain, of linked parts, nucleotides. There are four types: adenine (A), cytosine (C), guanine (G), and thymine (T). Their ordering carries a "code" that can be deciphered, showing how the information is passed down the line. As is well known, another nucleic acid, RNA, ribonucleic acid, lines up against the DNA, copies the coded information, and then picks out amino acids, complex organic molecules. These amino acids are then ordered, and they in turn make proteins, the building blocks of cells.

This is a paradigmatic example of thinking being guided by the machine metaphor. Such thinking transfers over into more traditional biological arguments. Consider the weird-looking dinosaur *Stegosaurus*, a brute that lived in the Jurassic period about 150 million years ago. It was very large – about nine meters long (30 feet) – and weighed rather more than five metric tons. Yet it had a very small brain, the size of that of a dog (less than three ounces). It was a herbivore, probably eating twigs and foliage and the like – hundreds of pounds a day. Unsurprisingly, it was probably very slow, five miles an hour maximum. Very puzzling is a line of plates along its back (Fig. 1.10). Why does the Stego have these plates? What is their function? Could they be for sexual attraction? Probably not, because both males and females have them, unlike the peacock/peahen, for instance, where the males have magnificent tail feathers whereas the females do not. More popular is the hypothesis by de Buffrénil, Farlow, and de Ricqlès that they

(a)

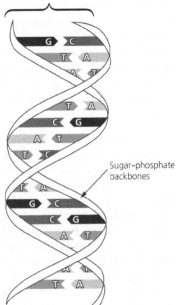

(b)

DNA double helix

Adenine **(A)**   Thymine **(T)**   Guanine **(G)**   Cytosine **(C)**

Figure 1.9   a, Nucleotide. **b**, Double helix.

Figure 1.10 *Stegosaurus.*

were for some kind of species recognition – attracting fellow Stegos and avoiding non-fellow Stegos.

Another hypothesis is the suggestion that the plates are for fighting, defense particularly. This is the hypothesis that is endorsed in a popular book on the dinosaurs by L. B. Halstead, published in 1975. Evidence? Apparently, new evidence suggests that the armour plates grew out sideways rather than upwards, and that meant that *Stegosaurus* was better positioned to strike out at enemies coming in from the side (Fig. 1.11). Unfortunately (for its originator), a year or two after this was written, his hypothesis all came tumbling down. It just wasn't convincing, because the plates do not grow out of the main skeleton, but are, as it were, add-ons. They simply would have been useless for fighting, attack or defense, because they would at once get ripped off. Such a postulated orientation of the plates, according to de Buffrénil, Farlow, and de Ricqlès, makes a testable prediction: to act as armor, the plates should be formed of

Figure 1.11 Halstead's view – Stegosaurus with plates sideways.

thick, compact "Panzer" bone, whereas there is histological evidence that their structure is extremely light and hollow.

A new, far more plausible hypothesis came on the scene. Like everything about the *Stegosaurus*, it is controversial; but, whether it turns out to be primary or not, serves beautifully as an example of how machine metaphor thinking is ubiquitous in the life sciences. This is the explanation that the function of those rather peculiar plates is to cool the brute in the height of a summer day. Remember, it is big and slow, and has need of massive amounts of food. Overheating is a real threat. The wind, blowing across the plates, brings down their temperature. And pertinent to our discussion, the reasoning behind this is totally machine-based. According to Farlow, Thompson, and Rosner, writing in 1976, "Wind tunnel experiments on finned models, internal heat conduction calculations, and direct observations of the morphology and internal structure of stegosaur plates support this hypothesis, demonstrating the comparative effectiveness of the plates as heat dissipaters controllable through input blood flow rate, temperature, and body orientation (with respect to wind)" (Fig. 1.12). Mechanisms all the way down! Mechanisms, but not to the exclusion of final cause. The "compact engineering devices" explain how the plates exist and work, *in order to* keep the animal at a functioning temperature. The animals would have been

Figure 1.12 Cooling tower fans.

living out in the open, on prairies or the like, where the winds blow and the plates are cooled. This would not have been the case had they lived in jungles or forests.

Relative value, the plates keep the brute cool; not absolute value, dinos are good things. Mechanism triumphant!

Organicism Redux

Except, of course, it wasn't. In Germany, at the end of the eighteenth century/ beginning of the nineteenth century, organicism was revitalized, starting a tradition that thrives (somewhat noisily) today. The German Romantics, *Naturphilosophen*, called for a replacement of "the concept of mechanism" and a renewal of the organic metaphor, "elevating it to the chief principle for interpreting nature," as historian Bob Richards has put it. Prominent names included the poet Johann Wolfgang von Goethe (Fig. 1.13), the anatomist Lorenz Oken, and, in respects the most influential of all, the philosopher Friedrich Schelling. Little surprise that from one who, as a teenager, wrote

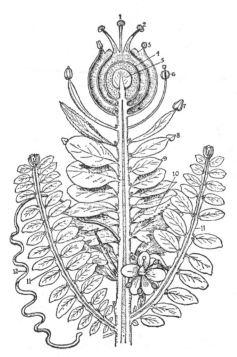

Figure 1.13   Goethe's *Urpflanze*, or archetypal plant. The individual organs of the plant, the lower simpler leaves, then the more complicated stem leaves, then the sepals, which are again designed quite differently, and the flower petals, which even have a different color than the stem leaves, are all externally different in form from each other, but inwardly in the idea are the same.

a 60-page essay on the *Timaeus*, it was he who most fervently pushed the Platonic vision of the whole world as an organism: "one power, one pulse, one life." Wanting, as an idealist, to break down the distinction between the objective – the world out there – and the subjective – the world in here, he found the answer in the Theory of Forms. "The key to the explanation of the entirety of the Platonic philosophy" he said, is "noticing that Plato everywhere carries the subjective over to the objective" (*History*, 212).

It was inevitable that Schelling rejected the Kantian judgment that the need to take account of the end-directedness of organisms spelt the second-rate nature of biological understanding. If one side of human awareness and understanding is the subjective, then at some level this must be reflected in the other side, the objective. In other words, if final-cause thinking is needed in our science, as it is, then in some real sense it must exist out in the world, to be discovered not created – which has the immediate implication that the *Timaeus* was right, the physical world must be essentially organic, subject to final causes as much as to efficient causes.

> Even in mere organized matter there is life, but a life of a more restricted kind. This idea is so old, and has hitherto persisted so constantly in the most varied forms, right up to the present day – (already in the most ancient times it was believed that the whole world was pervaded by an animating principle, called the world-soul, and the later period of Leibniz gave every plant its soul) – that one may very well surmise from the beginning that there must be some reason latent in the human mind itself for this natural belief. (*Ideas*, 35)

"Organized matter." The world's nature is such that it produces itself, its development comes from within, as an unfurling organism is produced by forces within rather than without. The acorn naturally grows up into the oak. From the simple to the complex, from the undifferentiated to the highly differentiated.

> Nature should be Mind made visible, Mind the invisible nature. Here then, in the absolute identity of Mind in us and Nature outside us, the problem of the possibility of a Nature external to us must be resolved. The final goal of our further research is, therefore, this idea of Nature; if we succeed in attaining this, we can also be certain to have dealt satisfactorily with that Problem. (42)

What then of humans? In a way not true of mechanism, organicism privileges humans. Darwin saw from the first that his theory did not support progress. There is no guarantee that humans will – must – exist: "there is no ≪NECESSARY≫ tendency in simple animals to become complicated" (*Notebooks*, 1836–1844, M 147). Natural selection is more than a tautology – those that survive are those that survive – but it is relativistic. When there is a lot of food available, there are adaptive advantages to being big. Little food available – there are adaptive advantages to being small. The same even applies to brains and intelligence. Little stability – there are advantages to having lots of offspring, and intelligence is a luxury. Lots of stability – few offspring, and intelligence is obligatory. There is no absolute value in being human. In the immortal words of the late paleontologist Jack Sepkoski (quoted in my *Monad to Man*): "I see intelligence as just one of a variety of adaptations among tetrapods for survival. Running fast in a herd while being as dumb as shit, I think, is a very good adaptation for survival" (486).

This is not at all the conclusion of the organicist. Values exist out there, objectively. Humans are the beings of greatest value. Hence the whole force of change – note that, in a way not true of mechanism, as Goethe and others recognized, evolution is built into organicism – is in the direction of ever greater worth, ending with the human species. Such thinking is the very essence of Romanticism, as Schelling put it: "It is One force, One interplay and weaving, One drive and impulsion to ever higher life" (*Proteus of Nature*). Unsurprisingly, as the nineteenth century progressed, in Germany particularly, trees of life were always topped by *Homo sapiens* (Fig. 1.14).

Going west, and crossing the Channel to Britain, the key figure is the general man of just about everything, Herbert Spencer, who appeared on the scene around 1850 and was still going strong 50 years later, as the old century died and the new century was born. As with Germanic Romantic holists, Spencer argued strongly in 1860 that societies are organisms, with interconnected parts, and saw this throughout the living world. For him, change comes from outside disruptions that take groups from stability and stasis and cause consequent change and upwards movement, from the homogeneous (all the same) to the heterogeneous (differences). Spencer called this in 1862 his theory of "dynamic equilibrium." Ultimately, what counts is the effort made by the individual. Outside factors trigger change but they do not cause it.

Expectedly, we find that Spencer finds value in the very process. For him, progress was the bedrock of all his thinking: "this law of organic progress is the law of all progress. Whether it be in the development of the Earth, in the

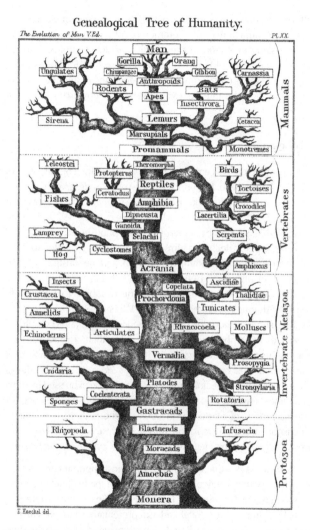

Figure 1.14 The tree of life as drawn by Ernst Haeckel in *The Evolution of Man* with humans at the top.

development of Life upon its surface, in the development of Society, of Government, of Manufactures, of Commerce, of Language, Literature, Science, Art, this same evolution of the simple into the complex, through successive differentiations, holds throughout" (*Progress*, 245). He explains that the English language is more complex and hence above all others. What greater proof could there be that, as evolution moves upwards, things are improved?

The Twentieth Century

In 1924, the Harvard faculty was enriched by the arrival of the English logician Alfred North Whitehead – who, with Bertrand Russell, was deservedly famous for the attempt (in their magnum opus *Principia Mathematica*) to show that mathematics follows deductively from the laws of logic. In the 1920s, moving into metaphysics, Whitehead gave a series of lectures, published as *Science and the Modern World*. Openly declaring himself an organicist, he called for "the abandonment of the traditional scientific materialism, and the substitution of an alternative doctrine of organism" (99). Continuing: "Nature exhibits itself as exemplifying a philosophy of the evolution of organisms subject to determinate conditions" (115). Although it seems likely that Whitehead got this second-hand, there is little surprise about the major influence here. "Nature should be Mind made visible, Mind the invisible nature. Here then, in the absolute identity of Mind in us and Nature outside us, the problem of the possibility of a Nature external to us must be resolved" (Schelling's *Ideas*, 42). Putting things together, Whitehead writes: "The final summary can only be expressed in terms of a group of antitheses, whose apparent self-contradictions depend on neglect of the diverse categories of existence. In each antithesis there is a shift of meaning which converts the opposition into a contrast." Sounds Hegelian, which is hardly a surprise given Hegel's deep roots in German Romanticism.

A case can be made for saying that Whitehead is the most important organicist – certainly the most important Anglophone organicist – of the twentieth century. No need for us to make such a judgment now. We shall (in Chapter 3) return to the topic. Here, let's simply take Whitehead as representative and, with an eye to our interests, move straight to the present. There is no great surprise in finding that prominent philosophers continue to work within the organicist tradition.

John Dupré stresses the special nature of humans – a nature of great worth given our unique abilities. Humans uniquely are capable of thinking, meaning we uniquely can see the whole and make judgments and act on them. He stresses that he thinks only humans have genuine freedom, at the same time denying that our superior nature puts any pressure on our animal origins.

Dupré is not alone in having organicist yearnings. Fellow philosopher Thomas Nagel offers explicit confirmation of the suspicion that we are now dealing with non-mechanical laws and understanding. The very title of his book – *Mind and Cosmos: Why the Materialist Neo-Darwinian Conception of Nature Is Almost Certainly False* – prepares the way. For Nagel, the adaptive nature of the organic world is far too complex for so crude a mechanism of natural selection. He suspects that "there are natural teleological laws governing the development of organization over time, in addition to laws of the familiar kind governing the behavior of the elements." He agrees that this takes us back to an Aristotelian view of nature alien to modern science. However, somewhat defiantly, he defends his position. Modern science may rule it out. Thomas Nagel does not (22). Nagel prides himself on being a non-believer, an atheist. Yet as an Aristotelian he must believe in some kind of absolute value. A question to be raised (later) is whether one can have such value in the absence of a deity.

And with that question left dangling, let us bring this chapter to an end.

2 The Mechanists' God

Warfare?

Evolution was both known and generally accepted very rapidly after the *Origin*. Darwinism, even. In the summer of 1860, Charles Dickens' widely read weekly, *All the Year Round*, carried two anonymously published articles on the theme of the *Origin* (with a third early in the next year).

> Man, by selection in the breeds of his domestic animals and the seedlings of his horticultural productions, can certainly effect great results, and can adapt organic beings to his own uses, through the accumulation of slight but useful variations given to him by the hand of Nature. But Natural Selection is a power incessantly ready for action, and is as immeasurably superior to man's feeble efforts, as the works of Nature are to those of Art. Natural Selection, therefore, according to Mr. Darwin – not independent creations – is the method through which the Author of Nature has elaborated the providential fitness of His works to themselves and to all surrounding circumstances. (174)

The author, David Thomas Ansted, a professional geologist and long-time acquaintance of Darwin, stressed that the religious need have little worry. Darwin is a serious scientist and a man of impeccable social and moral worth. Ostensibly agnostic about Darwin's work, the author leaves little doubt about his own positive position. Opponents are "timid":

> We are no longer to look at an organic being as a savage looks at a ship—as at something wholly beyond his comprehension; we are to regard every production of nature as one which has had a history;

we are to contemplate every complex structure and instinct as the summing up of many contrivances, each useful to the possessor, nearly in the same way as when we look at any great mechanical invention as the summing up of the labour, the experience, the reason, and even the blunders, of numerous workmen. (*Natural Selection*, 299)

Most revealing is that in 1850, honors degrees in the sciences were finally introduced at the University of Cambridge. One final exam in biology asked candidates to write an essay on why evolution is false. In the final exam of 1865, candidates were told to assume the truth of evolution and to get on and discuss causes. That, at the end of the decade, Charles Darwin's son Frank got a first shows obviously that being a Darwinian was no handicap.

What's the place of religion in all of this? The popular metaphor for the relationship between science and religion after the *Origin* is that of "warfare." It is epitomized by the clash in the summer of 1860, at the British Association for the Advancement of Science's annual meeting at Oxford *Wilberforce and Huxley*, (Lucas, 1979). Champion of the Church Bishop Samuel ("Soapy Sam") Wilberforce of Oxford (son of William Wilberforce of slave-trade-abolition fame) squared off against Darwin's champion, Professor (of natural history at the Royal School of Mines) Thomas Henry Huxley (Fig. 2.1). Supposedly, Wilberforce asked Huxley if he was descended from monkeys on his grandfather's side or his grandmother's side. Supposedly, Huxley responded that he had rather been descended from a miserable ape than from a bishop of the Church of England. A fine time was had by all, although one suspects that this is one of those stories where the fictional one – at least, elaborated-on one – is more fun, and indeed more revealing, than the true one. Certainly, real or apparent, it seemed to give the definitive account of the science–religion relationship in the age of evolution. Backed by two highly tendentious books claiming that science and religion are fated forever to be at war – J. W. Draper's *History of the Conflict Between Religion and Science*, published in 1875, and President of Cornell A. D. White's *History of the Warfare of Science with Theology in Christendom*, published in 1896 – that seemed to be the end of that.

Yet there has to be more to the story. Consider words penned towards the end of the nineteenth century, by the High Anglican Oxford academic Aubrey Moore. "Darwinism appeared, and, under the guise of a foe, did the work of a friend." He continues that we must choose:

> Either God is everywhere present in nature, or He is nowhere. He cannot be here, and not there. He cannot delegate his power to demigods called "second causes." In nature everything must be His work or nothing. We must frankly return to the Christian view of direct Divine agency, the immanence of Divine power from end to end, the belief in a God in Whom not only we, but all things have their being, or we must banish him altogether. (*Christian Doctrine*)

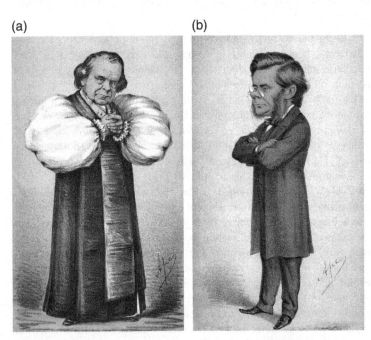

Figure 2.1   **a**, Samuel Wilberforce by Carlo Pellegrini, as he appeared in *Vanity Fair* (1869). **b**, Thomas Huxley by Carlo Pellegrini, as he appeared in *Vanity Fair* (1871).

Obviously not everyone thought that the relationship between science and religion, evolutionary thinking and Christianity, was one of conflict. Some apparently thought there was a reinforcing harmony.

I will begin this chapter with a little history, showing why the warfare/conflict thesis had such a hold on people's imagination and why in respects this hold continues. The story will tell us that there is no need for us to be trapped into thinking that this is the beginning and end of discussion. Then I shall turn to look at the work of those in Aubrey Moore's tradition, that is, those who think that there is perhaps even something constructive to come out of the science–religion interaction. In this chapter, my focus is on the mechanist perspective, and this means that Darwinian evolutionary theory is the science that will be foremost. In the next chapter I shall turn to the organicist perspective. I start with God and in later chapters will discuss what one might call subsidiary or entailed claims by/for the Christian.

Creationism

We have seen that those who claim that a Christianity based on the Bible, taken literally, is traditional Christianity are simply wrong. St Augustine was very clear on that. The Bible is true, totally, unequivocally. But often it needs to be read metaphorically or allegorically. Those who worry endlessly about how God could have created light on the First Day but waited until the Fourth Day to create the sun and the moon are simply wasting their time. What is crucial for the Christian is that God is Creator of everything. He did this freely from nothing, and it is good. Also, as becomes clear, it is the abode prepared for human beings.

By the beginning of the nineteenth century, this kind of thinking was widely accepted, especially among the educated. Geology is showing that the supposed creation in six days, six thousand years ago, is simply false. It goes against the physical evidence of the rocks. Scorn was heaped upon Oxford geologist William Buckland when, in the 1820s, he claimed that a cave in the East Riding of Yorkshire yielded evidence of Noah's Flood. If you look at William Whewell's discussion in his *History of the Inductive Sciences* (1837) as to why evolution cannot be true, the Bible doesn't have a dog in the fight. What counts is natural theology, design in particular. If this be so, why then even today – especially today – do we have very vocal biblical literalists? The answer lies, as do so many other rather strange offshoots of Christianity – Latter-day Saints, Seventh-day Adventists, Christian Scientists, Christadelphians, Jehovah's Witnesses – in the United States in the nineteenth century.

The roots of indigenous biblical literalism lie in the southern states, together with the new lands as the country began its push west. As is known to anyone who has read the *Little House on the Prairie* series – far better as books than the gooey television series – it was a harsh and unfriendly country that was being settled: from droughts and winter storms to dangerous animals, not to mention not-always-friendly humans long there and now being disturbed, as well as the distances and efforts needed to socialize and live normal community life – stores, schools, churches. Itinerant preachers played crucial roles, with crusades and other ways of spreading the gospel. Thanks to the industrialization of printing, books – Bibles – were much more freely available and affordable. It was natural that the Holy Book started to take a very prominent place – what one should believe, what one should do. What are the proper relationships between man and wife? How does one train and discipline one's children? Servants and slaves? What is owed to them? What is expected of them? Look to the Bible for advice.

The newly founded Seventh-day Adventists had a major role in structuring the ways in which the Bible was approached and read. They were literalists, grounded in their need to take the six days of Creation as six periods of 24 hours. It was to be anticipated that this would get caught up in the disputes leading to the Civil War. Anti-slavers, in the North, cited the Bible – the Beatitudes for instance – as evidence against the owning and subjection of other human beings. In the South, to the contrary, the Bible was cited in favor of slavery. Ephesians 6:5–9 was a particular favorite:

> [5] Slaves, obey your earthly masters with respect and fear, and with sincerity of heart, just as you would obey Christ. [6] Obey them not only to win their favor when their eye is on you, but as slaves of Christ, doing the will of God from your heart. [7] Serve wholeheartedly, as if you were serving the Lord, not people, [8] because you know that the Lord will reward each one for whatever good they do, whether they are slave or free.

If St Paul said this, who is to argue otherwise?

After the Civil War the Bible, taken literally, continued to play a big role – in the South, in the newly occupied lands of the mid-West, and increasingly in poorer regions of large cities, where the already-existing inhabitants felt threatened by the flux of European immigrants – Catholic, with a leavening of Jews. For the South, the story of the Israelites in captivity to the Babylonians was much appreciated. God inflicts greatest hardships on those whom He loves most. This was enshrined in a series of pamphlets, the *Fundamentals*, published at the beginning of the twentieth century. Hence the popular term of "Fundamentalists," for biblical literalists. More recently people refer to themselves as "Creationists" (or "Scientific Creationists").

Expectedly, opposition to evolutionary theorizing (especially Darwinian evolutionary theorizing) played a major role in the thinking of Creationists. The Bible, taken literally, talks of six days of Creation, of a single pair as the ancestors of all subsequent humans, of a worldwide flood, and much more. All of this is denied by Darwinism. However, it was not until after the First World War that things really came to a head when a schoolteacher in Tennessee, John Thomas Scopes, was put on trial for teaching evolution. Prosecuted by three-times presidential candidate William Jennings Bryan and defended by notorious secular thinker Clarence Darrow, the trial became a public spectacle, widely reported in America and abroad (Fig. 2.2). Scopes was found guilty, although, on a technicality, his conviction was overturned on appeal. Nevertheless, the trial had a chilling effect on what was permissible in the classroom. As always, sales figures rated more highly than dissemination of the truth, and publishers of biology textbooks, seeing which way the wind was blowing, dropped all mention of Darwin and his theory – until events like Sputnik, in 1957, the success of which was taken as a sad confirmation of the inadequacy of American science education. New Darwin-friendly texts were commissioned and distributed by the Biological Sciences Curriculum Study program, notably *Biological Science: Molecules to Man*. The evangelicals found that their children were being taught the hated evolution – as gospel, to use a phrase. There was immediate reaction. In 1961, a biblical scholar and a hydraulic

Figure 2.2 Clarence Darrow and William Jennings Bryan, 1925.

engineer, John C. Whitcomb and Henry M. Morris, published *Genesis Flood*, arguing against evolution and for the literal truth of the whole Holy Bible – six-day creation, universal deluge, parting of the Red Sea, and so on down to the present. It was hugely successful – the emphasis on the Flood, rather than Creation, reflected the fear of atomic conflict and the feeling that it was a foreteller of worse to come, Armageddon. Through the 1960s and 1970s, Creationism gained strength, and things came to a climax in 1981, when the State of Arkansas passed a law insisting that in biology classes of publicly funded schools of the state, Creationism (or Creation Science) be given "balanced treatment" along with the teaching of evolution.

At once, the American Civil Liberties Union sprang into action, arguing that the law violates the First Amendment of the Constitution, because it egregiously mixes Church and State. Another trial ensued – along with popular scientist Stephen Jay Gould and leading Protestant theologian Langdon Gilkey, I (a philosopher) was a witness for the prosecution. There was little surprise that the judge, William Overton, ruled firmly against Creationism. He said flatly:

> More precisely, the essential characteristics of science are:
>
> (1)   It is guided by natural law;
> (2)   It has to be explanatory by reference to natural law;
> (3)   It is testable against the empirical world;
> (4)   Its conclusions are tentative, i.e. are not necessarily the final word; and
> (5)   It is falsifiable. (Ruse and other science witnesses.)
>
> Creation science ... fails to meet these essential characteristics.

Evolutionary thinking is science. Creationism is religion and, as such, can have no place in the classrooms of publicly funded schools.

Intelligent Design Theory

Creationism had apparently met its Armageddon. Yet, as always, things were never quite this easy. The Creationists quit trying to do things through legislation. Rather, they worked (and very successfully) at nagging away at individual school boards and the like, getting textbooks banned from classes and so forth. As importantly, they developed a smoother version of their position, Creationism-lite, which they called "Intelligent Design Theory." Now the emphasis was less on the Bible taken literally and more on the need of an intelligence – more precisely, Intelligence – to get all started and going. Much was made of the notion of what was called "irreducible complexity," supposedly to be shown by biological structures such as the bacterial flagellum (Fig. 2.3). The idea was that such structures could not have been produced by a gradual process such as evolution because, for their functioning, they require all parts in place at the same time. Remove one, and nothing works. Leading exponent of IDT Michael Behe wrote: "By irreducibly complex I mean a single system composed of several well-matched, interacting parts that contribute to the basic function, wherein the removal of any one of the parts causes the system to effectively stop functioning" (*Darwin's Black Box*, 39).

Well, what about the bacterial flagellum? The flagellum is a kind of tail, driven by a rotary motor, to move the bacterium. Every part is incredibly complex,

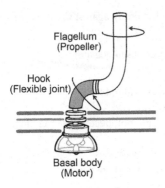

Figure 2.3 Bacterial flagellum.

and so are the various parts combined. The external filament of the flagellum (called a "flagellin"), for instance, is a single protein that makes a kind of paddle surface contacting the liquid during swimming. Near the surface of the cell, just as needed, is a thickening, so that the filament can be connected to the rotor drive. This naturally requires a connector, known as a "hook protein." There is no motor in the filament, so that must be somewhere else. "Experiments have demonstrated that it is located at the base of the flagellum, where electron microscopy shows several ring structures occur" (70). All way too complex to have come into being in a gradual fashion. Only a one-step process will do. Darwinism is ruled out and we must look for another explanation. There are two options – a "hopeful monster" after the style of Lucretius, or an "intelligent agent," after the style of Paley (96). There is only one possible answer. Irreducible complexity spells design.

What is the response? Critics claim that Behe shows a misunderstanding of the very nature and workings of natural selection. No one is denying that, in natural processes, there may well be parts which, if removed, would lead at once to the non-functioning of the systems in which they occur. The point, however, is not whether the parts now in place could not be removed without collapse, but whether they could have been put in place by natural selection. Consider an arched bridge, made from cut stone, without cement, held in place only by the force of the stones against each other. If you tried to build the

Figure 2.4 Keystone bridge.

bridge from scratch, upwards and then inwards, you would fail – the stones would keep falling to the ground, as indeed the whole bridge now would collapse were you to remove the center keystone or any surrounding it (Fig. 2.4). Rather, what you must do is first build a supporting structure (possibly an earthen embankment), on which you will lay the stones of the bridge, until they are all in place. At this point you can remove the structure, for it is no longer needed, and in fact is in the way. Likewise, one can imagine a biochemical sequential process with several stages, on the parts of which other processes piggyback, as it were. Then the hitherto non-sequential para-sitic processes link up and start functioning independently, the original sequence finally being removed by natural selection as redundant or incon-veniently draining of resources.

This is all pretend. But Darwinian evolutionists have hardly ignored the matter of complex processes. Indeed, it is discussed in detail by Darwin in the *Origin*, where he refers to that most puzzling of all adaptations, the eye. At the biochemical level, today's Darwinians have many examples of the most complex of processes that have been put in place by selection. Take that staple of the body's biochemistry, the process where energy from food is converted into a form which can be used by the cells. Rightly do textbooks refer to this vital organic system, the so-called "Krebs cycle," as something which "undergoes a very complicated series of reactions." This process, which occurs in the cell parts known as mitochondria, involves the production of ATP (adenosine triphosphate): a complex molecule which is energy-rich and which is degraded by the body as needed (say in muscle action) into another less rich molecule, ADP (adenosine diphosphate). The Krebs cycle remakes ATP from other energy sources – an adult human male needs nearly 200 kg a day – and by any measure, the cycle is enormously involved and intricate. For a start, nearly a dozen enzymes (substances that facilitate chemical processes) are required, as one sub-process leads on to another (Fig. 2.5).

Yet the cycle did not come out of nowhere. It was cobbled together out of other cellular processes that do other things. It was a "bricolage." Each one of the bits and pieces of the cycle exists for other purposes and has been coopted for the new end. Meléndez-Hevia and colleagues, the scientists who have made this connection, could not have made a stronger case against Behe's irreducible complexity than if they had had him in mind from the first. In fact, they set up the problem virtually in Behe's terms: "The Krebs cycle has been frequently quoted as a key problem in the evolution of living cells, hard to explain by Darwin's natural selection: How could natural selection explain the building of a complicated structure in toto, when the intermediate stages have no obvious fitness functionality?" What these workers do not offer is a Behe-type answer. First, they brush away a false lead. Could it be that we have something like the evolution of the mammalian eye, where primitive existent eyes in other organisms suggest that selection can and does work on proto models (as it were), refining features that have the same function, if not as efficient as more sophisticated models? Probably not, for there is no evidence of anything like this. But then we are put on a more promising track. Consider the Krebs cycle, the method by which cells generate energy. Every part, every essential part, started life doing something else. There was no mind, real or apparent, planning the whole from the start. The Krebs cycle came about through a process that François Jacob, in 1977, called "evolution by molecular tinkering." Evolution doesn't make new things from scratch. It

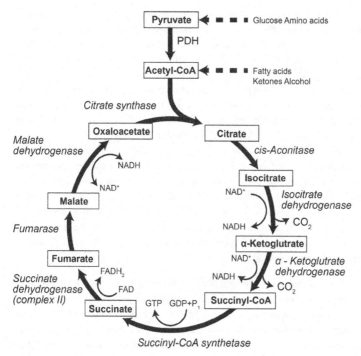

Figure 2.5 Krebs cycle.

fiddles around with what already exists. Jacob concludes: "a chemical engineer who was looking for the best design of the process could not have found a better design than the cycle which works in living cells" (302). "Chemical engineer"!

Rounding off the response to Behe, let us note that, if his arguments are well founded, then in some respects we are in bigger trouble than otherwise. His position seems simply not viable given what we know of the nature of mutation and the stability of biological systems over time. When exactly is the intelligent designer supposed to strike and to do its work? In Behe's *Darwin's Black Box*, the suggestion is made that everything might have been done long ago and then left to its own devices. "The irreducibly complex biochemical systems that I have discussed ... did not have to be produced recently. It is entirely possible, based simply on an examination of the systems themselves, that they were designed billions of years ago and that they have been passed down to the present by the normal processes of cellular reproduction." Unfortunately, as Behe's most doughty biological critic retorts, we cannot ignore the history of the preformed genes between the point of their origin (when they would not have been needed) and today when they are in full use. Ken Miller explains: "As any student of biology will tell you, because those genes are not expressed, natural selection would not be able to weed out genetic mistakes. Mutations would accumulate in these genes at breathtaking rates, rendering them hopelessly changed and inoperative hundreds of millions of years before Behe says that they will be needed." There are masses of experimental evidence showing that this is the case. Behe's idea of a designer doing everything back then and then leaving matters to their natural fate is "pure and simple fantasy."

What is the alternative? Presumably that the designer is at work all the time, producing mechanisms as and when needed. If we are lucky, we might expect to see some produced in our lifetime. Indeed, there must be a sense of disappointment among biologists that no such creative acts have so far been reported. More than this, as we turn from science towards theology, there are even greater disappointments. Most obviously, what about mal-mutations? If the designer is needed and available for complex engineering problems, why could not the designer take some time on the simple matters, specifically those simple matters that if unfixed lead to horrendous problems, redolent of some of the problems detailed by David Hume in his *Dialogues Concerning Natural Religion* (1779). Some of the worst genetic diseases are caused by one little alteration in one little part of the DNA. Sickle-cell anemia, for example. If the designer is able and willing to do the very complex because it is very good, why does it not do the very simple because the alternative is very bad? Behe speaks of this as being part of the problem of evil, which is true but not very helpful. Given that the opportunity and ability to do good was so obvious and yet not taken, we need to know the reason why.

Revealed Religion

Move on. We now know that most if not all the warfare/conflict thesis is based on bad biology and bad Christianity. Let us now steer around to a position, if not identical to that of Aubrey Moore, within the same domain. To do this, we first remind ourselves of the traditional – Augustinian – conception of God. God is Creator. Outside space: "No physical entity existed before heaven and earth." Outside time: "Your Today is eternity." How do we get to know and believe in this God? Traditionally, there are two ways: revealed religion or natural theology. The first gets at God through faith. The second gets at God through reason and evidence. Taking them in order, let us start with revealed religion.

Although, as a junior Quaker, I was brought up on the *Screwtape Letters*, I cannot stand C. S. Lewis' *Narnia* series. Wardrobes are just not my thing. But I understand fully (even although I have never shared the experience) when he writes of his conversion to Christianity (*Surprised by Joy*). Reason and evidence had nothing to do with it. Rather, it was his sense that God was reaching out to him.

> You must picture me alone in that room in Magdalen [College, Oxford], night after night, feeling, whenever my mind lifted even for a second from my work, the steady, unrelenting approach of Him whom I so earnestly desired not to meet. That which I greatly feared had at last come upon me. In the Trinity Term of 1929 I gave in, and admitted that God was God, and knelt and prayed: perhaps, that night, the most dejected and reluctant convert in all England.

Theologically, this makes perfect sense. It is precisely what one would expect of God. Belief in God through faith is what the Calvinist philosopher Alvin Plantinga calls "properly basic": "the believer is entirely within his intellectual rights in believing as he does even if he doesn't know of any good theistic argument (deductive or inductive), even if he believes there isn't any such argument, and even if in fact no such argument exists." Calvin wrote in 1536 of the *sensus divinitatis*.

> That there exists in the human minds and indeed by natural instinct, some sense of Deity, we hold to be beyond dispute, since God himself, to prevent any man from pretending ignorance, has endued all men with

some idea of his Godhead, the memory of which he constantly renews and occasionally enlarges, that all to a man being aware that there is a God, and that he is their Maker, may be condemned by their own conscience when they neither worship him nor consecrate their lives to his service. (*Institutes* 1, 3, 1)

Generations of philosophy students, whose only knowledge of the philosophy of religion comes through books of readings, having encountered Aquinas' five proofs for the existence of God through extracts from the *Summa Theologica* taken entirely out of context, unsurprisingly tend to assume that the great theologian/philosophers rated natural theology above revealed religion. Reason and evidence above faith, which latter is really a bit second-rate. Nothing could be further from the truth. The discussion starts with Jesus and the disciple "Doubting" Thomas, who refuses without proof to acknowledge Jesus as the risen lord. "Then saith he to Thomas, Reach hither thy finger, and behold my hands; and reach hither thy hand, and thrust it into my side: and be not faithless, but believing. And Thomas answered and said unto him, My Lord and my God. Jesus saith unto him, Thomas, because thou hast seen me, thou hast believed: blessed are they that have not seen, and yet have believed" (John 20:27–9).

All the great theologians have endorsed this at one level or another. Aquinas thought that reason can sometimes get there in the end: "For certain things that are true about God wholly surpass the capability of human reason, for instance that God is three and one: while there are certain things to which even natural reason can attain, for instance that God is, that God is one, and others like these" (*Summa Theologica* 1259–1265, 5). Note, however, that reason – where we could be wrong – is limited, and in the end, faith – where we cannot be wrong – is top dog. "The truth of the intelligible things of God is twofold, one to which the inquiry of reason can attain, the other which surpasses the whole range of human reason" (7). Ultimately, without faith you only get part of the story, and Aquinas makes clear that faith trumps all – how else could the ignorant or stupid or lazy get knowledge of God? John Paul II, in his encyclical *Fides et Ratio*, affirmed this position strongly: "The results of reasoning may in fact be true, but these results acquire their true meaning only if they are set within the larger horizon of faith: 'All man's steps are ordered by the Lord: how then can man understand his own ways?' [Proverbs 20:24]."

As is well known, since the nineteenth century, one strand of thought has stressed that, in some sense, even trying to prioritize natural theology is a mistake. Danish theologian Søren Kierkegaard stressed the fundamental priority of faith, as have theologians in the twentieth century, notably the Swiss theologian Karl Barth. Their basic position was not only that faith took priority but that in some way using evidence and reason to support it – or to play an equal role – was to debase faith, to undermine it. Faith is only faith if it requires a leap – a "leap into the absurd." You have to be prepared to take a chance, to make a commitment. Kierkegaard explains:

> When someone is to leap he must certainly do it alone and also be alone in properly understanding that it is an impossibility ... the leap is the *decision*. ... But if a resolution is required, presuppositionlessness is abandoned. The beginning can occur only when reflection is stopped, and reflection can be stopped only by something else, and this something else is something altogether different from the *logical*, since it is a resolution.

And:

> Faith is the objective uncertainty with the repulsion of the absurd, held fast in the passion of inwardness, which is the relation of inwardness intensified to its highest. This formula fits only the one who has faith, no one else, not even a lover, or an enthusiast, or a thinker, but solely and only the one who has faith, who relates himself to the absolute paradox. (*Concluding Unscientific Postscript*)

Atheists don't care for much of this. It is escapism at best, pernicious nonsense at worst. New Atheist Chicago biologist Jerry Coyne writes:

> The danger to science is how faith warps the public understanding of science: by arguing, for instance, that science is based just as strongly on faith as is religion; by claiming that revelation or the guidance of ancient books is just as reliable a guide to the truth about our universe, as are the tools of science; by thinking that an adequate explanation can be based

on what is personally appealing rather than what stands the test of empirical study. (*Faith versus Fact*, 225–6)

Expectedly, the non-believer can give all sorts of psychological reasons for faith, starting with cowardice in the face of the unknown, the unknown after death particularly. But this is not going to trouble the person of faith, who might well accept the psychology! The point is that, as a Christian, one expects God to speak to one directly – and He does! As expectedly, for Coyne this is not enough. "My claim is this: science and religion are incompatible because they have different methods for getting knowledge about reality, have different ways of assessing the reliability of that knowledge, and, in the end, arrive at conflicting conclusions about the universe."

The person of faith will respond that Christianity is simply not in the business of rivaling or replacing science. This was the mistake of the Creationists. It is rather than religion and science are independent, offering different answers to different questions. Take the story of Noah and the Flood. For the post-Augustinian Christian, this is not a science story, a story about geology. It is rather a moral fable. People were misbehaving, so, other than Noah and family, God wiped them all out. And what happens when the waters subside, and they get back out onto dry land? Noah gets stinking drunk, and one of his kids laughs at him in his nakedness. Misbehavior! In short, the story is really about the futility of simplistic solutions. Would that George W. Bush and Tony Blair had taken note of this before they invaded Iraq.

It sounds rather like the solution being urged here is akin to Stephen Jay Gould's notion of different "Magisteria" – different kinds of world picture answering different kinds of questions. This is true as far as it goes, but Gould then goes on to say that science is about the real world whereas religion is about morality. However, although the religious person undoubtedly would agree that religion is about morality – the interpretation just given about Noah's Flood is moral – contra Gould, the religious person would surely want to assert that religion is about more than morality. It makes factual claims. However, this is not the thin end of a large wedge being hammered in by Jerry Coyne. The kinds of factual claim made by the person of faith are not those of science. This follows from science falling within the machine metaphor. "Why is there something rather than nothing?" This is what Martin

Heidegger called the "fundamental question" of metaphysics. Science has no answer. You might mention the Big Bang, but why the Big Bang? Was there anything before it? Machines are like those hares in the cookbook. First take your materials, and then see what you can produce with them. Religion does have an answer: a good necessary being, God, created it all out of love. A factual claim but not a scientific claim.

There is more one could say on this topic. We can and will pick up the discussion later. Here, we have enough to see why the person of faith is going to be supremely indifferent to the kind of charges made by those such as Jerry Coyne. What about his last-gasp criticism? "'Knowledge' acquired by religion is at odds not only with scientific knowledge, but also with knowledge professed by other religions. In the end, religion's methods, unlike those of science, are useless for understanding reality." John Hick has the answer: "What would one expect of humans immersed in culture, as are we?" Little surprise that Jesus of Nazareth did not give the Sermon on the Mount to people in Tibet. They simply would not have understood it. The point is that the ultimate is a mystery. Hick explains:

> Let us begin with the recognition, which is made in all the main religious traditions, that the ultimate divine reality is infinite and as such transcends the grasp of the human mind. God, to use our Christian term, is infinite. He is not a thing, a part of the universe, existing alongside other things; nor is he a being falling under a certain kind. And therefore, he cannot be defined or encompassed by human thought. We cannot draw boundaries around his nature and say he is this and no more. If we could fully define God, describing his inner being and his outer limits, this would not be God. The God whom our minds can penetrate and whom our thoughts can circumnavigate is merely a finite and partial image of God.

Little surprise, either, at Hick's thinking. He was drawn to Quakerism, and that religion has always taken a mystical approach to the Godhead. In the words of the population geneticist, J. B. S. Haldane: "My own suspicion is that the universe is not only queerer than we suppose, but queerer than we *can* suppose." For Hick, all religions have elements of truth, but only elements, because they are trying to map the Unknown and Unknowable. Hick's theology is "apophatic." We cannot say what God is. We can say only what God is not.

Natural Theology

Turn now to getting at God through reason and evidence. There are a number of traditional arguments for the existence of God. One of the most famous is the so-called "ontological argument," first formulated by St Anselm and later taken up by René Descartes in his *Meditations*. This is an a priori argument, deriving God's existence and nature from His very definition. If, in Anselm's language, God is by definition "that than which none greater can be thought" – in Descartes' language, the being with "all perfections" – then necessarily He exists and is good. Why? For Anselm, it is because otherwise one could think of a being which is better because It exists; for Descartes, because otherwise one could think of a being with at least one more perfection, the perfection of existence! Few philosophers have been taken in by this argument. As Kant pointed out, there is something a bit shifty about claiming that existence is a perfection. However, it is almost the mark of a philosopher to admire the audacious nature of the argument, trying to get so much from so little. Of importance to us here is that as an a priori argument the ontological argument really doesn't impinge on the science–religion debate.

The same is true of other arguments, for instance the causal or cosmo-logical argument, which tries to infer the existence of God from the fact that things of our world demand causes, and the only possible ultimate cause is that which has itself no need of cause, that which exists neces-sarily, namely God. Of course, you might say that the whole point of science is that it demands causes, to which the counter is that, when we introduce God, we are going beyond science. However, as already intim-ated in Chapter 1, the teleological argument or the argument from design – organisms look design-like, hence there is a Designer (for Plato) or some kind of special vital "final cause" law (for Aristotle) – does come up against the life sciences, Darwin's process of natural selection in particu-lar. In order to get adaptations, final-cause-type entities, all you need are things to follow the paths dictated by blind laws. Does this actually point us away from God? As we saw, in his deistic mode, in the *Origin* Darwin rather suggested it did the opposite. "To my mind it accords better with what we know of the laws impressed on matter by the Creator, that the production and extinction of the past and present inhabitants of the world should have been due to secondary causes ... ." Richard Dawkins is hardly likely to go this far, but he too seems to allow that some form of religious belief may be possible: "Although atheism might have been logically tenable before Darwin, Darwin made it possible to be an intellectually fulfilled atheist." It leaves God standing if that is important to you. God is down on His knees if not flat on His back for the skeptic or non-believer.

Yet it does seem that, on reflection, Darwin became a little more hardline. He grew to agree with those who argued that it was the theist's God or no God at all. At the end of the large, two-volume work he published later in 1868 – *Animals and Plants under Domestication* – Darwin argued that he could not believe in a God who created all sorts of variations and, knowing the winner, stepped back and let nature run its course. Suppose "an architect were to rear a noble and commodious edifice, without the use of cut stone, by selecting from the fragments at the base of a precipice wedge-formed stones for his arches, elongated stones for his lintels, and flat stones for his roof" (see Fig. 2.4). Would we truly want to say that the stones were not random, because God had been behind making them as they are? Darwin continues:

> The shape of the fragments of stone at the base of our precipice may be called accidental, but this is not strictly correct; for the shape of each depends on a long sequence of events, all obeying natural laws; on the nature of the rock, on the lines of deposition or cleavage, on the form of the mountain which depends on its upheaval and subsequent denudation, and lastly on the storm or earthquake which threw down the fragments. But in regard to the use to which the fragments may be put, their shape may be strictly said to be accidental.

Isn't this a case of out of the frying pan and into the fire? One comes up against a major difficulty.

> An omniscient Creator must have foreseen every consequence which results from the laws imposed by Him. But can it be reasonably maintained that the Creator intentionally ordered, if we use the words in any ordinary sense, that certain fragments of rock should assume certain shapes so that the builder might erect his edifice?

Darwin is arguing against Asa Gray, who claimed that the variations used by selection are guided by God. Darwin thought (truly) that this makes selection redundant. He came to see that the variations produced by a deistic God are no less directed than the variations produced by a theistic God. The deistic God makes the variations needed, doing this deliberately. He then just covers this up by creating a slew of variations with no use. In the end, there is no real difference between the successful variations of theist and deist. Either way, natural selection is redundant. The argument from design needs guided variations. Natural selection makes guided variations unnecessary. No need to assume a Designer – indeed, the existence of a Designer seems precluded. Hence, natural selection destroys the argument from design, even as it keeps the major premise, that organisms are as-if designed. Thus, in his *Autobiography* (written around 1875) Darwin felt able to write: "The old argument of design in nature, as given by Paley, which formerly seemed to me so conclusive, fails, now that the law of natural selection has been discovered."

Agnosticism

But should one be an atheist? That seems to be jumping the gun a little. If Darwin shows you don't have to be a believer, wouldn't it be best to opt for agnosticism – neither belief in God nor disbelief in God? This is the position of Thomas Henry Huxley, he who invented the term "agnostic." None of the traditional categories – theism, deism, atheism – seem to fit his own position. Referring to these people, Huxley wrote:

> The one thing in which most of these good people were agreed was the one thing in which I differed from them. They were quite sure they had attained a certain "gnosis," – had, more or less successfully, solved the problem of existence; while I was quite sure I had not, and had a pretty strong conviction that the problem was insoluble. And, with Hume and Kant on my side, I could not think myself presumptuous in holding fast by that opinion. So I took thought, and invented what I conceived to be the appropriate title of "agnostic." It came into my head as suggestively antithetic to the "gnostic" of Church history, who professed to know so

much about the very things of which I was ignorant. To my great satisfaction the term took. (*Evolution and Ethics*)

We see, for Huxley, the God question brought out his uncompromising unwillingness to accept an easy or comforting position. As the not-altogether-friendly critic Thomas Spencer Baynes said truly, there was something of a Calvinist about Huxley. "He has the moral earnestness, the volitional energy, the absolute conviction in his own opinions, the desire and determination to impress them upon all mankind, which are the essential characteristics of the Puritan character," continuing: "His whole temper and spirit is essentially dogmatic of the Presbyterian or Independent type, and he might fairly be described as a Roundhead who had lost his faith. He himself shows the truest instinct of this in calling his republished essays 'Lay Sermons'" (*Darwin on Expression*, 502).

In Huxley's world, the pursuit of truth through science was a deep moral obligation, however cold and unfriendly it may seem.

> What I earnestly maintain is, that thoroughly good work in science cannot be done by any man who is deficient in high moral qualities –
>
> It is the moral which is essential to the right working of the intellect – and the value of science is that it compels men to know that such is the case. (Letter to F. Dyster, January 30, 1859)

Huxley was a determined mechanist who saw no values in science itself, the world revealed by science. Values are ours, or at least our subjective response to eternal verities. Thus, when, in his late essay *Evolution and Ethics*, he saw that the struggle for existence is the secret to human evolution, he stressed vigorously that this was not a moral conclusion. Rather, it is for us to apply morality showing that the struggle for existence does not point the way to right behavior. In other words, science might – does – point to a "cold and unfriendly world," but this has no implicit value. Huxley would have agreed with Thomas Hardy, in his poem "Hap," that the world is indifferent to our needs and thoughts.

> —Crass Casualty obstructs the sun and rain,
> And dicing Time for gladness casts a moan...
> These purblind Doomsters had as readily strown
> Blisses about my pilgrimage as pain.

Huxley's agnosticism is that morally we do and should do science, but there is neither justification nor demolition underpinning it. Indeed, he invented the term "agnostic" precisely because it was neutral, unlike the popular "humanism," which he saw as a religion about humans. Relatedly, he could never have been an atheist because they – like Richard Dawkins today – were, in his opinion, also in the religion business.

Fine-tuning

Is this the end of discussion? On the positive side, is there still a chance of getting to God through evidence and reason? Why should one go beyond John Henry Newman, who (after the *Origin*) wrote in a letter: "I believe in design because I believe in God, not in God because I believe in design" (*Letters*, 97). Why should one not extend this line of thinking to another argument often trumpeted as proving God's existence – the only explanation of miracles is the existence of a good God? But why should one not argue: I believe in miracles because I believe in God, not in God because I believe in miracles? Indeed, one might surely say that a lot of the miracles were not miraculous in the sense of demanding a break in law. The marriage at Cana, water into wine, is better explained as Jesus shaming the host enough that he brought up his best, hitherto-hidden wine. That is surely a miracle, as is getting those with food to share with the five thousand. Even the Resurrection is explained, not as a conjurer's trick, like sawing the lady into half, but as, on that fateful Sunday, the disciples sitting around totally rejected and despairing, suddenly feeling in their hearts: "My Redeemer Liveth." What a miracle, even if a psychologist might have a natural explanation in terms of group hysteria or whatever! Miracles are about meaning, not about metaphysics.

Before we give up, we must acknowledge a group of enthusiasts today who think the design argument can be retrieved. It focuses on the notion of so-called "fine-tuning." This relies on what is known as the "anthropic principle": the laws of nature cannot be due to chance, because if the laws were even fractionally different from the way that they actually are, then no life could have been produced, let alone flourish. "If the initial explosion of the big bang had differed in strength by as little as one part in 10^{60}, the universe would have either quickly collapsed back on itself, or expanded too rapidly for stars to form. In either case, life would be impossible. As John Jefferson Davis points out, an accuracy of one part in 10^{60} can be compared to firing a bullet at a one-inch target on the other side of the observable universe, twenty billion light years away, and hitting the target." Following up, Robin Collins put it: "if gravity had been stronger or weaker by one part in 10^{40}, then life-sustaining stars like the sun could not exist. This would most likely make life impossible." And from this is drawn the inference that it cannot be chance. The universe is "fine-tuned," pointing to Something Higher.

Put to one side the surely legitimate worry that, however successful the fine-tuning argument may be, it does not have a lot to do with the life sciences. If it works, then perhaps we should not be so complacent about its being irrelevant to the life sciences. Looking at the argument in its own right, part of the problem here is that one really doesn't know the options. We have no way of experimenting and have only our universe to judge from. Think of a number, double it, and the answer you want is a half. While it is certainly true that on our planet the only life we know is carbon-based, dare one say – can one offer a kind of ontological argument in reverse – that there is no other possible form of life, a form that could do all that is theologically necessary? Would God love us any the less if we were made of silicon rather than carbon? Moreover, what if there are multiverses, universes parallel to ours, possibly having all sorts of different laws and constants? In other words, our universe is one of very many. Of course, ours is a universe that works, in the sense that it can and did produce life. But this is no big issue. We wouldn't be in it if it didn't work. Only one person in a million is going to win the lottery, but there is no miracle about the person who is the winner being the winner. Someone had to win. Perhaps no universe had to produce life, but our existence shows that life could be produced and, if you have enough universes and combinations, then our existence seems no more miraculous than that someone won the lottery.

Even if there are no multiverses, one is still working blind. The physics Nobel Laureate Steven Weinberg is not impressed. He notes that one of the favorite examples of supposed fine-tuning is the carbon atom. This is something that did not occur in the early moments of the universe. Back then, everything was just hydrogen and helium. It had to be formed, and it seems that for carbon we need three helium nuclei. However, normally this cannot happen because the energy of carbon is way below that of three helium nuclei. Fortuitously, there is a radioactive form of carbon that has just the higher energy that is needed, and so everything works out just fine. But before you dash in and say that it is not just fine but fine-tuned, keep digging. The three helium nuclei come together in a two-part process. First two of them combine to make beryllium and then the third is added to make carbon. It turns out here that there is significantly more wiggle room, that it is the energy level at this level that is crucial for the production of carbon, and that in fact there is a range of possible energies that would do the job. In short, perhaps the laws we have were not so tightly designed.

The Problem of Evil

Fine-tuning? Nice try, but no cigar. What about going the other way? Are there natural theological arguments disproving the existence of God? Darwin worried about this one, writing just after the *Origin* to his American, Christian friend, Asa Gray:

> I had no intention to write atheistically. But I own that I cannot see, as plainly as others do, & as I sh^d wish to do, evidence of design & benefi-cence on all sides of us. There seems to me too much misery in the world. I cannot persuade myself that a beneficent & omnipotent God would have designedly created the Ichneumonidæ with the express intention of their feeding within the living bodies of caterpillars, or that a cat should play with mice. Not believing this, I see no necessity in the belief that the eye was expressly designed. (Letter to Asa Gray, May 22, 1860)

The problem of evil! How can one believe in an all-powerful, all-loving God when there is so much pain and suffering in the world? The Lisbon earthquake? Auschwitz? Childhood leukemia?

It is usual to divide the problem into two: moral evil – Auschwitz – and natural or physical evil – the Lisbon earthquake. As far as natural evil is concerned, Richard Dawkins of all people has a powerful argument. With few exceptions – Descartes in his *Meditations* being one – it is agreed that

God cannot do the impossible. God cannot make $2 + 2 = 5$. To have functioning life, you must have a process that produces adaptation, as well as a liveable environment. Processes like Lamarckism or saltationism – evolution by lucky monsters – simply don't work. It is natural selection or nothing. But if natural selection, then you must let the world run according to unbroken law with its consequences – the struggle for existence, for example. You are bound to get pain and strife. Something like childhood leukemia is just the luck of the draw. You produce complex machines like humans and you are going to have failures, malfunctions. And to produce sunshine and fertile fields and so forth, you must pay the cost of earthquakes and the like. I am not sure how appreciative Dawkins would be to learn that he did not get to this point first; it was Aquinas: "lions would not thrive unless asses were killed" (*Summa Theologica* 1a, 25, 6). Some – emphatically not Richard Dawkins! – would combine this kind of thinking with the view that this struggle is a good thing, for we come out better in the end. Popular here is the sentiment of the poet John Keats. "The common cognomen of this world among the misguided and superstitious is 'a vale of tears' from which we are to be redeemed by a certain arbitrary interposition of God and taken to Heaven – What a little circumscribed straightened notion! Call the world if you please 'The vale of soul-making'" (Letter, 1819). Thanks to suffering, we become stronger human beings.

What of moral evil? The usual God-exonerating argument is that it is a function of free will. Better to let people sin than to make them robots. If the latter, they do not have the potential to be good or bad, a consequence God most certainly did not want. Of course, you must now accept that the free will of Heinrich Himmler was equal to the worth of six million Jews; more modestly, of equal worth to one Anne Frank. But, as the saying goes, you cannot make an omelette without cracking eggs. Does Darwinian evolution, mechanistic evolution, have anything to say about this? A powerful distinction is made between r-selection and K-selection. The former, r-selection, is good in times of fluctuation and instability. A good reproductive strategy is to have lots of offspring, even if the consequence is little parental care. Herrings, for instance. When times are good, you are ready to take advantage of them, and in bad times you are not much worse off. The latter, K-selection, is good in times of stability. Few offspring and lots of parental care, demanding decisions. Then you are maximizing your chances of success. Humans! So in a sense, one might say that some dimension of freedom is the virtue of – or more pessimistically – the tragedy of having evolved with large brains able to make choices. How this applies within our species is, as they say, a much-contested topic. We will leave more detailed discussion of the free will problem until later (Chapter 4). The topic is being postponed, not ignored.

Humans Contingent?

Bringing to an end our discussion of natural theology in the post-Darwinian era, there is one more question about the necessity of human beings. It is surely a central claim of Christianity that humans must be here on Earth. You can make Genesis as allegorical as you like, it remains the case that it privileges humans. We had to exist. If the best that evolution can do is dinosaurs, Christianity is in trouble. We, and we alone, are made in the image of God. Our intelligence; our moral sense. But does evolutionary theory, mechanistic Darwinian evolutionary theory, support the claim that it is a necessary consequence that humans will in fact exist? In other words, does evolution claim that there is progress towards humans?

As we have seen, when Darwin thought consciously about the matter, he could see no progress. No guarantee about humans. Yet Darwin, like everyone else, believed in social progress, and he saw evidence of it in nature. He was, after all, a well-heeled member of the British Empire at its zenith. From the *Origin*: "The inhabitants of each successive period in the world's history have beaten their predecessors in the race for life, and are, in so far, higher in the scale of nature; and this may account for that vague yet ill-defined sentiment, felt by many palæontologists, that organisation on the whole has progressed" (267). And there is, of course, the notorious passage at the end of the *Origin* talking of the power of selection:

> Thus, from the war of nature, from famine and death, the most exalted object which we are capable of conceiving, namely, the production of the higher animals, directly follows. There is grandeur in this view of life, with its several powers, having been originally breathed into a few forms or into one; and that, whilst this planet has gone cycling on according to

the fixed law of gravity, from so simple a beginning endless forms most beautiful and most wonderful have been, and are being, evolved. (490)

Yet, in the end, within the mechanist paradigm, this progress is contingent, not, as in the organicist paradigm, a necessity, a given.

Arms Races

Not to say that this has stopped people from trying to find a way to generate progress within the mechanist paradigm; value without necessarily presupposing a deity. A particularly interesting move made by Julian Huxley, in his first book, *The Individual in the Animal Kingdom*, published in 1912, was to try to offer a kind of neo-Darwinian mechanism for organic progress, one of the keystones of the organicist program, but now shown to be justifiable in mechanistic terms! He offered a cultural–biological analogy, using the turn-of-the-century naval arms race between Britain and Germany as the example. First, the cultural: "Halfway through the century, when guns had doubled and trebled their projectile capacity, up sprang the 'Merrimac' and the 'Monitor,' secure in their iron breast-plates; and so the duel has gone on"; concluding: "Each advance in attack has brought forth, as if by magic, a corresponding advance in defence." Then, the biological: "With life it has been the same: if one species happens to vary in the direction of greater independence, the inter-related equilibrium is upset, and cannot be restored until a number of competing species have either given way to the increased pressure and become extinct, or else have answered pressure with pressure" (115).

As it happens, in the third edition of the *Origin*, Darwin had already given an interesting anticipation of the arms-race metaphor.

> If we look at the differentiation and specialisation of the several organs of each being when adult (and this will include the advancement of the brain for intellectual purposes) as the best standard of highness of organisation, natural selection clearly leads towards highness; for all physiologists admit that the specialisation of organs, inasmuch as they perform in this state their functions better, is an advantage to each being; and hence the accumulation of variations tending towards specialisation is within the scope of natural selection. (134)

He then qualifies it all away!

> On the other hand, we can see, bearing in mind that all organic beings are striving to increase at a high ratio and to seize on every ill-occupied place in the economy of nature, that it is quite possible for natural selection gradually to fit an organic being to a situation in which several organs would be superfluous and useless: in such cases there might be retrogression in the scale of organisation. (134)

In 1872, in a letter to the American biologist Alphaeus Hyatt, Darwin wrote:

> Permit me to add that I have never been so foolish as to imagine that I have succeeded in doing more than to lay down some of the broad outlines of the origin of species. After long reflection I cannot avoid the conviction that no innate tendency to progressive development exists, as is now held by so many able naturalists, & perhaps by yourself. It is curious how seldom writers define what they mean by progressive development; but this is a point which I have briefly discussed in the Origin.

In the same year as this letter, in the sixth edition of the *Origin*, Darwin wrote: "natural selection, or the survival of the fittest, does not necessarily include progressive development – it only takes advantage of such variations as arise and are beneficial to each creature under its complex relations of life" (98). One can hardly say that Darwin makes an overwhelming case for the Christians' need of a guarantee that human beings will appear on Earth once life has started. For all the lure of progressive evolution, he knew that once he had introduced any kind of inevitable direction to evolution, that was curtains for natural selection.

Notwithstanding Darwin's doubts, later thinkers, notably Richard Dawkins, have picked up on Darwin's arms-race metaphor. Dawkins too finds the key in arms races. As one who embraced computer technology early and enthusiastically, perhaps expectedly Dawkins notes that, more and more, today's arms races rely on computer technology rather than brute power, and – in the animal world – he finds this translated into ever bigger and more efficient brains. No need to hold your breath about who has won. Dawkins invokes a notion known as an animal's EQ, standing for "encephalization quotient." This is a kind of cross-species measure of IQ that takes into account the amount of brain power needed simply to get an organism to function (whales require much bigger brains than shrews because they need more computing power to get their bigger bodies to function), and that then scales according to the surplus left over. Dawkins writes in *The Blind Watchmaker*: "The fact that humans have an EQ of 7 and hippos an EQ of 0.3 may not literally mean that humans are 23 times as clever as hippos! But the EQ as measured is probably telling us something about how much 'computing power' an animal probably has in its head, over and above the irreducible amount of computing power needed for the routine running of its large or small body" (189).

As always, it is the analogy with human progress that is the key. At a conference in Melbu (Norway) in 1989, Dawkins stressed how the development of computers involved a kind of hardware/software coevolution. Advances in hardware were symbiotically connected with advances in software. Apparently, there is also software/software coevolution – you make advances in software in one area or dimension, and this leads to advances in software in other areas or dimensions. He adds that he was trying to "suggest, by my analogy of software/software coevolution, in brain evolution that these may have been advances that will come under the heading of the evolution of evolvability in the evolution of intelligence."

Does any of this do what is needed? Even if there is something to what Dawkins writes, it seems questionable whether advances in military hardware are always going to be a metaphor for advances in intelligence, or progressive for that matter. Although it seems to have been highly effective, one doesn't see much progress in the use of poison gas in the First World War – at least not progress in any sense that one would identify as morally worthwhile (which is presumably what one wants for humans). Same for atomic weapons in the Second World War. Although computers today are hugely more efficient than they were even a few decades ago, I am not sure we always get the kinds of progress that is a good analogy for human beings, or, if they are, then the conclusion seems to be that humans are not at all the kinds of beings that are supposed by Christians. Christians, particularly those of the Augustinian tradition, stress human sinfulness; yet the underlying premise is that we are the creation of a good God and hence are essentially good. It is just that we have gone astray. But would one want to say that computers are essentially good? The whole point is that – in this mechanical world – computers are inherently value-free and can be used for good and for bad indifferently. No doubt there will be a range of opinions as to which of these categories this laptop-written book falls into.

Go back to multiverses. If there are an infinite number, then presumably since humans could evolve – we have evolved – somewhere at some time humans were bound to evolve. The trouble, of course, is that presumably an infinite number of humans could have evolved and did evolve. Not to mention all the failures, with the IQ of a turnip and the sporting ability of Michael Ruse. One much doubts that Darwin would have been impressed. Although he obviously had no thoughts of multiverses, one suspects that, in the light of his argument at the end of his *Animals and Plants under Domestication*, he would have wanted to rule even that option off limits. A God who knew the answer before He began, because He had set things up to get the answer He wanted, was a cheat. Not worthy of His status. Not worthy of our worship.

Enough. For now, as we prepare to turn to the organicist approach or approaches to the God question, let us sum up and ask where the mechanist – the Darwinian – leaves the question. Revealed religion escapes virtually unscathed, mainly because (with good reason) it refuses to get into a fight. Natural theology is more complex. Perhaps T. H. Huxley was right. The traditional proofs do not establish the existence of the Christian god. Christianity is not proven, but neither is atheism.

3 The Organicists' God

Organicism and Christianity

Ask the basic question: Can a Christian be an organicist? And respond with the basic answer: Yes! There are Christians who welcome the idea of an organic Earth, at the least. Thomas Berry, a Catholic priest no less, had a theological vision of the world that made the organic thesis central. "The universe is not a vast smudge of matter, some jellylike substance extended indefinitely in space. Nor is the universe a collection of unrelated particles. The universe is, rather, a vast multiplicity of individual realities with both qualitative and quantitative differences, all in the spiritual-physical community with one another" (*Sacred Universe*, 71–2). Integration pushes us to the Earth-as-an-organism. "This unique mode of Earth-being is expressed primarily in the number and diversity of living forms that exist on Earth, living forms so integral to one another and with the structure and functioning of the planet that we can appropriately speak of Earth as a 'Living Planet'" (110). What about a bit of Darwin bashing? "Darwin had only a minimal awareness of the cooperative and mutual dependence of each form of life on the other forms of life. This is remarkable: he himself discovered the great web of life, yet he did not have a full appreciation of the principle of intercommunion" (73). One infers that Berry was an evolutionist but of the more directed, monad to man, version.

These days, one is not quite sure where the British Royal Family stand on Christianity. For all that the monarch is the head of the Church of England, one does truly think that King Charles III is closer to Paganism than he is to the Thirty-Nine Articles. He is an enthusiastic organicist. Not only does he talk to plants – makes sense, one supposes, if we are all siblings in this business together – but he also quotes the alchemic "Emerald Tablet of Hermes": "And as all things are One, so all things have their birth from this One Thing by adaptation. Its power is integrating if it be turned into Earth" (*Harmony*, 120). One suspects that the king's subjects will be able to bear such outré thinking with equanimity. However, I do not think one is being unduly prejudicial in not expecting similar sentiments from members of the Church of Jesus Christ of Latter-day Saints – the Mormons. The Church is rather conservative – the surprise would be were Senator Mitt Romney not a Republican. Certainly, were his beliefs otherwise, one would not expect him to have been elected. Utah is not the place one goes to find tree huggers. Try California. As it happens, however, whatever the senator's private beliefs or inclinations, he would be out of line were he to eschew organicism publicly. It is an integral part of the theology of the Latter-day Saints. We learn from *The Book of Moses* (dictated to Joseph Smith in 1830 and 1831, and now incorporated in *The Pearl of Great Price*, one of the four sacred books of the Mormon Canon) that the Earth itself is quite able to express fairly strong emotions: "Wo, wo is me, the mother of men; I am brained, I am weary, because of the wickedness of my children. When shall I crest, and be cleansed from the filthiness which is gone forth out of me? When will my Creator sanctify me, that I may rest, and righteousness for a season abide upon my face?" (7.48). The talk is of a Creator, but the world is more now than just of value. It is a conscious being of value.

If one does start to take seriously the idea of the world – the whole universe – as in some sense organic, then this starts to raise the question of consciousness. Obviously, if one thinks of the world as a plant – acorn to oak – this is a little less pressing. But as one veers towards thinking of the world as an animal, the question becomes to what extent it can control its own destiny, let alone our destiny. Joseph Smith in the passage above talks of a Creator, but he also imbues the world with a conscious mind of some sort. Sanctify, rest, righteousness. These are not terms one associates with oak trees. How one answers this question is clearly in major part dependent on one's philosophy of mind. The mechanist does not run into this problem. Not only do machines have no inherent value, they don't think, either. Leibniz made that clear. You are working in the Cartesian world of material substance, *res extensa*. No consciousness in. No consciousness out.

> In imagining that there is a machine whose construction would enable it to think, to sense, and to have perception, one could conceive it enlarged while retaining the same proportions, so that one could enter into it, just like into a windmill. Supposing this, one should, when visiting within it, find only parts pushing one another, and never anything by which to explain a perception. Thus it is in the simple substance, and not in the composite or in the machine, that one must look for perception. (*Monadology*, 215)

This is not at all to say that the mechanist can say nothing about the mind. You can talk about the parts of the brain and how they serve different functions. But for the rest, we seem stuck in Haldane-like ignorance: the world is not only queerer than we think it is, it is queerer than we could think it is. Unless, of course, one starts to move beyond what we might call stripped-down mechanism. Interestingly, since virtually the time of the *Origin*, it has been suggested that Darwinian evolutionary theory might help here. Since evolution is gradual, instead of looking for consciousness to appear, suddenly, fully functioning, we should look for something gradual about consciousness. William Kingdom Clifford wrote:

> The only thing that we can come to, if we accept the doctrine of evolution at all, is that even in the very lowest organism, even in the Amoeba which swims about in our own blood, there is something or other, inconceivably simple to us, which is of the same nature with our own consciousness, although not of the same complexity. (*Body and Mind*, 38–9)

This position is known philosophically as "panpsychic monism." Material and mental are one. This means that, if you are committed to the world as an organism, you already believe that in some sense the world is conscious. One suspects that many believers find this philosophy congenial. The question is whether it is a philosophy congenial to Christianity. At the most immediate level, one is drawn to pantheism. As Spinoza said: *Deus sive Natura* – God or Nature. The trouble is that one comes up against an unscalable roadblock. For the Christian, the world and its inhabitants are not God. They are the children of God. It is a well-established truth that, by the time they are 30, children seem to turn into their parents. My daughter Emily had to major in Afternoon Studies, because she could not get up in the morning. Now, says a proud father, in her thirties she is a public defender, down at the city jail at eight o'clock on a Saturday morning, dealing with the miscreants from the night before. One would like to think she is like her dad, although I cannot honestly remember ever giving a class on the synthetic *a priori* at eight on a Saturday morning, but neither of us would want to say that she is her dad. That is what the Christian says. We are made in the image of God. We are not God. So much so, says the Augustinian, that we needed God to die on the Cross for our salvation. We could not do it on our own.

Our conclusion thus far is that the Christian can certainly be an organicist. With respect to evolution, organicism is particularly attractive because it – like Christianity – privileges humans. It also finds value in the world, as does the Christian. "And God made the beast of the earth after his kind, and cattle after their kind, and every thing that creepeth upon the earth after his kind: and God saw that it was good" (Genesis 1:25). Remember, however, all the value comes from God. It is not self-created. This is why, presumably, one can be a panpsychic monist, but one must be careful not to identify this with pantheism. Universal consciousness is a creation of God, not God Himself.

Henri Bergson

Take this discussion a little further. How does the Christian organicist see God working His purpose in the world? In the French "vitalist" philosopher Henri Bergson – who posited a kind of Aristotelian life force, the *élan vital*, to explain the upward course of evolution – we see an attempt to solve, or transcend, this difficulty. Formally, we do not see a great deal of traditional religion in Bergson's thinking, although at the end of his life – living in Vichy France – he did admit to feeling close to Catholicism, and would have converted, were it not that he did not want to evade the persecution of his fellow Jews. Bergson is not too keen on a teleological system as such. It seems to have a kind of determined finality about it. The acorn grows into the oak tree – it seeks and attains its end. No choice.

But radical finalism is quite as unacceptable, and for the same reason. The doctrine of teleology, in its extreme form, as we find it in Leibniz for example, implies that things and beings merely realize a programme previously arranged. But if there is nothing unforeseen, no invention or creation in the universe, time is useless again. As in the mechanistic hypothesis, here again it is supposed that *all is given*. Finalism thus understood is only inverted mechanism. (*Creative Evolution*, 39).

Nevertheless, Bergson does think there is something end-directed about everything. It is here that the *élan vital* plays its crucial role. It is this that lies behind evolution: "an *original impetus* of life, passing from one generation of germs to the following generation of germs through the developed organisms which bridge the interval between the generations. This impetus, sustained right along the lines of evolution among which it gets divided, is the fundamental cause of variations, at least of those that are regularly passed on, that accumulate and create new species" (87). It leads to a creative process. "If now we are asked why and how it is implied therein, we reply that life is, more than anything else, a tendency to act on inert matter. The direction of this action is not predetermined; hence the unforeseeable variety of forms which life, in evolving, sows along its path. But this action always presents, to some extent, the character of contingency; it implies at least a rudiment of choice" (96). Most particularly, we get the evolution of humans. "So that, in the last analysis, man might be considered the reason for the existence of the entire organization of life on our planet" (185).

We are certainly not getting a conventional picture of Christianity. Yet, obviously, Bergson felt he was sufficiently close that he could contemplate joining the Catholic Church, asking for prayers after his death. Most immediately, we think of the vital impetus, the *élan vital*, as in some sense representative of the Godhead. It is not material, it works on – and hence is separate from – our physical world. Above all, it is creative – it has free choice – and that is exercised in creating the truly superior, truly unique species: humankind, the apotheosis of the creative evolution. There are still a lot of questions, but we are edging closer to a more familiar theological world picture, a conclusion confirmed by the reply

Bergson penned to a (Catholic priest) critic in 1912. He stressed that for him God was the source of all, independent of His creation, which was done by him freely. Hence, the material is dependent on God, not God on the material. He adds that he is committed to "the idea of a God, creator and free, the generator of both Matter and Life, whose work of creation is continued on the side of Life by the evolution of species and the building up of human personalities. From all this emerges a refutation of monism and of pantheism" (*Bergson and His Philosophy*).

Omega Point

Continue the line of French organicist evolutionists. Listen to Sir Peter Medawar, discoverer of fundamental bodily processes opening the way to organ transplantation, and winner of the Nobel Prize, writing a review of Pierre Teilhard's *The Phenomenon of Man* in the leading philosophical journal *Mind*:

> It is a book widely held to be of the utmost profundity and significance; it created something like a sensation upon its publication a few years ago in France, and some reviewers hereabouts have called it the Book of the Year – one, the Book of the Century. Yet the greater part of it, I shall show, is nonsense, tricked out by a variety of tedious metaphysical conceits, and its author can be excused of dishonesty only on the grounds that before deceiving others he has taken great pains to deceive himself. *The Phenomenon of Man* cannot be read without a feeling of suffocation, a gasping and flailing around for sense.

That this outburst is right at the heart of the mechanism–organicism divide is made very clear at the beginning of the next paragraph of Medawar's review. "*The Phenomenon of Man* stands square in the tradition of *Naturphilosophie*, a philosophical indoor pastime of German origin which does not seem even by accident (though there is a great deal of it) to have contributed anything of permanent value to the storehouse of human thought."

Medawar embraced the tradition of Nobel Prize winners who assume that, because they have said it, they must be right. He was a molecular biologist, and he shows the contempt of a reductionist biologist for another field. "In his lay capacity Teilhard, a naturalist, practised a comparatively humble and unexacting kind of science." So much for paleontology! One can see why it raised Medawar's ire. To do its work, it not only uses the organicist's favorite concept, homology, but, because it deals with evolution through time, from the blob up to and including humans, it has tendencies towards progressionism. The English translation of Teilhard's *Phenomenon of Man* has an enthusiastic endorsement by Julian Huxley, and one suspects that, as much as Teilhard, Huxley is at the center of Medawar's sights. "Teilhard's belief, enthusiastically shared by Sir Julian Huxley, that evolution flouts or foils the second law of thermodynamics is based on a confusion of thought; and the idea that evolution has a main track or privileged axis is unsupported by scientific evidence." Huxley was president of the British Teilhard de Chardin society, and this is a good point to turn to Teilhard himself, for Bergson proves to be a major influence, as (one might have inferred) he was for Huxley.

> If this book is to be properly understood, it must be read not as a work on metaphysics, still less as a sort of theological essay, but purely and simply as a scientific treatise. The title itself indicates that. This book deals with man solely as a phenomenon; but it also deals with the whole phenomenon of man. (*Phenomenon of Man*, 28)

Those are the opening lines of *The Phenomenon of Man*. If ever they offer a prize for the most misleading opening of any book, it will be a heavy favorite. Teilhard de Chardin's book may be many things. Purely and simply a work of science it is not, for all that Medawar treated it – very negatively – as science. What then is *The Phenomenon of Man* if not science? Theology? Not in the eyes of Teilhard's own order, the Jesuits, nor indeed of the Catholic Church as a whole! A mere five years after its publication, we are told that the work abounds "in such ambiguities and indeed even serious errors, as to offend Catholic doctrine." The warning follows: "For this reason, the most eminent and most revered Fathers of the Holy Office exhort all Ordinaries as well as the superiors of Religious institutes, rectors of seminaries and presidents of universities, effectively to protect the minds, particularly of the youth, against the dangers presented by the works of Fr. Teilhard de Chardin and of his followers" (O'Connell, referring to the Monitum (Warning) by the Holy Office, 1962).

So Teilhard is anathematized in both the secular and the sacred world – one senses there must be something of interest here, especially when one learns from (secular) historian Jean Gayon that Teilhard was certainly the most brilliant French paleontologist of the first part of the twentieth century. On the side of religion, more recently, both the last pope and the present pope have written appreciatively of Teilhard's thinking. In this revisionist spirit, if we criticize Teilhard for misleadingly calling his book "science," let us also praise him for going beyond dialogue, science and religion simply talking to each other. He offered what was, at the time, the most systematic and audacious attempt at integration of evolutionary theory and the Christian religion.

The basic thesis of *The Phenomenon of Man*, written in the 1930s although only published posthumously in 1955, is quite simple. The world and life within it reveal an ongoing dynamic, a process of evolution of ever greater complexity, from the inorganic through the simplest organisms, up through various stages of existence to the highest, the "noösphere." This is the domain of humankind, culminating in something Teilhard called the Omega Point:

> Our picture is of mankind labouring under the impulsion of an obscure instinct, so as to break out through its narrow point of emergence and submerge the earth; of thought becoming number so as to conquer all habitable space, taking precedence over all other forms of life; of mind, in other words, deploying and convoluting the layers of the noosphere. This effort at multiplication and organic expansion is, for him who can see, the summing up and final expression of human pre-history and history, from the earliest beginnings down to the present day. (*Phenomenon of Man*, 190)

Inventively, if very controversially, Teilhard identified the climax, the Omega Point, with God as incarnated in Jesus Christ: "The universe fulfilling itself in a synthesis of centres in perfect conformity with the laws of union. God, the Centre of centres. In that final vision the Christian dogma culminates" (293). This, we learn, perfectly coincides with the Omega Point.

Inventive, but, overall, not entirely unfamiliar. In his Preface to *The Phenomenon of Man*, Julian Huxley writes that, before being ordained priest in 1912, Teilhard's reading of Bergson's *Evolution Creatrice* inspired a profound interest in the general facts and theories of evolution. The vision of evolution that is the backbone of *The Phenomenon of Man* is Bergsonian through and through – above all, organicist, in seeing ever greater complexity, driving towards some higher end: humans! If you come at Teilhard through the lens of a Darwinian mechanist, you may have science on your side, but you miss what he is about.

Does Teilhard then have no worries about running together what others might consider the separate domains of science and religion? Apparently not. *The Phenomenon of Man*, written by a Christian, does not linger over Heidegger's fundamental question: "Why is there something rather than nothing?" The answer is a given. God did it. Teilhard writes: "Traced as far as possible in the direction of their origins, the last fibres of the human aggregate are lost to view and are merged in our eyes with the very stuff of the universe" (36). At once, Teilhard makes it clear that he is thinking in terms of the nature of the already-existing stuff, and he excuses himself of even thinking much about this, because it is more a matter for physics than for his own field of biology. It is all a bit technical for the non-expert. "As I am a naturalist rather than a physicist, obviously I shall avoid dealing at length with or placing undue reliance upon these complicated and fragile edifices" (38).

Values do interest Teilhard. As a through and through organicist, he sees ever greater value the closer we get to humankind, with the corollary that the more humankind grows in complexity, the more value we have and the more value we want to generate: "conquered by the sense of the earth and human sense, hatred and internecine struggles will have disappeared in the ever warmer radiance of Omega. Some sort of unanimity will reign over the entire mass of the noösphere. The final convergence will take place in peace" (287). Panpsychism is the natural philosophy of consciousness for one who thinks in this mode: "we are logically forced to assume the existence in rudimentary form (in a microscopic, i.e. an infinitely diffuse, state) of some sort of psyche in every corpuscle, even in those (the mega-molecules and below) whose complexity is of such a low or modest order as to render it (the psyche) imperceptible – just as the physicist assumes and can calculate those changes of mass (utterly imperceptible to direct observation) occasioned by slow movement" (301).

And, thus to purpose and ends, that which gives the full (and only) meaning to Teilhard's world system. With the noösphere we are still going hand in hand with the secular organicist; but then, with the Omega Point, we go beyond. All is put in a Christian context: "The end of the world: the overthrow of equilibrium, detaching the mind, fulfilled at last, from its material matrix, so that it will henceforth rest with all its weight on God-Omega. The end of the world: critical point simultaneously of emergence and emersion, of maturation and escape" (287). Taken all in all, a wonderful world vision, even if it must be said: *C'est magnifique, mais ce n'est pas la science.*

Agnosticism

An important linking theme for all these organicist Christians is that their concept of God is traditional. He is the Augustinian being: Creator, outside time and space. That means their take on the revealed religion/natural theology debate is going to be conventional. As for all Christians, faith is primary. The distinctive thing, separating them from mechanists, is that, to a person, they think that natural theology works. The argument from design works. That comes with the territory. Indeed, this is a major reason why these people are organicists rather than mechanists. Are there going to be Christians who are not organicists? One supposes so. The population geneticist Ronald A. Fisher was a life-long member of the Church of England. Yet his take on evolution, his Darwinian take on evolution, was machine metaphor through and through. Most particularly, he devised what he called the "fundamental theorem of natural selection" about how selection works in populations: "the fundamental theorem proved above bears some remarkable resemblances to the second law of thermodynamics." Immodestly, though entirely typically, Fisher tells the reader that the second law is the supreme law of physics. "It is not a little instructive that so similar a law should hold the supreme position among the biological sciences" (*Genetical Theory*, 39).

Although a little less self-assured, Theodosius Dobzhansky was as important to evolutionary biology in the empirical sense as Fisher in the theoretical sense. He argued that there is always variation in natural populations, a claim that his student Richard Lewontin confirmed, and so selection does not have to wait for the unique right mutation. It always has other options. If a prey's camouflage is not right, then perhaps there is a variation for nasty taste, or going nocturnal, or . . . . Dobzhansky was a Christian, a member of the Russian Orthodox Church. He was also, amazingly, the president of the American branch of the Teilhard de Chardin Society – Teilhard who, as we have just seen, argued that evolution is progressive, up through humankind to the Omega Point, which he identified with Jesus Christ. Dobzhansky was openly a progressionist, seeing a process up to humans. This is from a letter to the historian of science John Greene, also a Christian, but not a Teilhardian: "Certain evolutionary processes are 'creative' because they bring about (a) something new (b) having an internal coherence since it maintains or advances life, and (c) may end in either success or failure. One of the notable successes, let us say the greatest success, was the origin of man." My sense is that Dobzhansky was not entirely clear how this happens. He was a Darwinian and wanted to stay in the machine metaphor world. But as a Christian, he was drawn to the organism metaphor world. Did he have any real choice?

Can one be a non-believing organicist? Remember Thomas Nagel, totally convinced of the coherence of the notion of teleological laws (*Mind and Cosmos*, 22)? Yet he is an atheist. Is his a coherent position? Can one be an Aristotelian organicist without an Unmoved Mover? Nagel is a panpsychist. Does his world consciousness do what is needed? Is he at a minimum being pushed to *Deus sive Natura*? The Platonic designer of external teleology is ruled out; but so also is the Unmoved Mover of internal teleology. If one nevertheless claims to be an Aristotelian, whence cometh the absolute value – or Value – towards which things are directed? If there is such a Value, what confers this status on it? Can you have Value without a valuer? And if the Value is described as something like Plato's Form of the Good, you are getting very close to being flipped back to Augustine's Christian God. Or to natural selection, a solution equally unpalatable to Nagel.

What of organicism and some form of skepticism? Like Thomas Henry Huxley, Herbert Spencer spoke of himself as an "agnostic"; but, expectedly, he was as far distant as it was possible to be from Huxley's stern secular Calvinism. In the sunny progressive world of Spencer, there is not going to be any warfare between science and religion. Serious thought

… must lead us to anticipate that the diverse forms of religious belief which have existed and which still exist, have all a basis in some ultimate fact. Judging by analogy the implication is, not that any one of them is altogether right; but that in each there is something right more or less disguised by other things wrong. It may be that the soul of truth contained in erroneous creeds is very unlike most, if not all, of its several embodiments; and indeed, if, as we have good reason to expect, it is much more abstract than any of them, its unlikeness necessarily follows. But however different from its concrete expressions, some essential verity must be looked for. (*First Principles* 1, 7)

It is when we try to pin down this verity that the trouble starts. Any attempt, for example, to explain why the world exists at all runs into paradox and incomprehensibility. We find out that "different suppositions respecting the origin of things, verbally intelligible though they are, and severally seeming to their respective adherents quite rational, turn out, when critically examined, to be literally unthinkable. It is not a question of probability, or credibility, but of conceivability" (2, 11). Little wonder that Spencer (who had Quaker influences) spoke of the "Unknowable." He is into (what we have already encountered as) "apophatic" theology. We can say what God is not. We cannot say what God is. Again, already identified as something akin to mysticism.

This view of religion is but half of what Spencer argued. When it comes to science, he thought we end up in precisely the same position as religion, ultimately facing the Unknowable. We do all the science we can, revealing as we know a value-impregnated (organicist) world; but, eventually, we come up against the same basic assumptions, about the cause of it all and about the underpinning beliefs pertaining to the world we are exploring. You can take gravity apart as much as you like, but, in the end, you are going to have to say: "I can go no further." The scientist "realizes with a special vividness the utter incomprehensibleness of the simplest fact, considered in itself. He, more than any other, truly knows that in its ultimate essence nothing can be known" (*First Principles* 3, 22). We end in the same position as with religion. Little wonder that there can be no essential conflict. You can speak of this as "agnostic" – Spencer does! But note how the organicist philosophy underpins it, as Huxley's mechanistic approach underpins his agnosticism. For Huxley, there is no searching for value, because it does not exist. The value comes from our own attitudes and behavior – continuing the search for knowledge, no matter what. For Spencer, there is something of value out there, even though we cannot grasp its nature. It is that which makes everything – the world, its inhabitants, humans – come into and stay in existence, in being. And the whole point of organicism is that there is value out there, to be found. And value increases in our quest for understanding, even though for us it will ever be incomplete. "That this is the vital element in all religions is further proved by the fact, that it is the element which not only survives every change, but grows more distinct the more highly the religion is developed" (2, 14). We start with primitive polytheistic belief systems. Then our thinking matures.

> The growth of a Monotheistic faith, accompanied as it is by a denial of those beliefs in which the divine nature is assimilated to the human in all its lower propensities, shows us a further step in the same direction; and however imperfectly this higher faith is at first realized, we yet see in altars "to the unknown and unknowable God," and in the worship of a God that cannot by any searching be found out, that there is a clearer recognition of the inscrutableness of creation. (*First Principles* 2, 14)

As always, Spencer's thinking is not only value-laden but essentially progressive, where progress tends to be spelt out in terms of Anglo-Saxon beliefs and society.

Process Philosophy

One final question. Thus far, the Augustinian God goes unscathed. His creation may be organic. He is not. Turn now to Alfred North Whitehead, the organicist's organicist. We have seen how, transplanted to Harvard in the 1920s, he became the leading, and highly influential, exponent of the "philosophy of organism." Above all, he regarded with horror the traditional God of Augustine. Surely influenced by the tragedy of the Great War – in which he and his wife lost their son Eric, aged 19, in the Royal Flying Corps – Whitehead turned his back on such an appallingly cruel creature. His God had to be down there in the trenches, facing the horrors with us. This is the God of the poet Pattiann Rogers.

> And maybe he suffers from the suffering
> Inherent to the transitory, feeling grief himself
> For the grief of shattered beaches, disembodied bones
> And claws, twisted squid, piles of ripped and tangled,
> Uprooted turtles and crowd rock crabs and Jonah crabs,
> Sand bugs, seaweed and kelp.
>
> (*Song of the World Becoming*, 182–3)

Whitehead and his followers wanted nothing to do with a God who feels no compassion for the family when they learn that their child has leukemia. Nothing to do with the God who is unmoved – cannot be moved because He is eternal and unchanging. "To sorrow, therefore, over the misery of others does not belong to God," to quote Aquinas (*Summa Theologica*). In any case, as an out-and-out follower of Schelling, on the one hand Whitehead took the inherent change of organicism as all-important, and, on the other hand, was totally committed to a God in the world rather than a God who is in some sense logically separate. Remember Schelling: "Nature should be Mind made visible, Mind the invisible nature. Here then, in the absolute identity of Mind in us and Nature outside us, the problem of the possibility of a Nature external to us must be resolved" (*Ideas*, 42). Whitehead writes:

> The vicious separation of the flux from the permanence leads to the concept of an entirely static God, with eminent reality, in relation to an entirely fluent world, with deficient reality. But if the opposites, static and fluent, have once been so explained as separately to characterize diverse actualities, the inter-play between the thing which is static and the things which are fluent involves contradiction at every step in its explanation. (*Process and Reality*, 346)

He tells us that "[t]he final summary can only be expressed in terms of a group of antitheses, whose apparent self-contradictions depend on neglect of the diverse categories of existence. In each antithesis there is a shift of meaning which converts the opposition into a contrast." Thus:

> It is as true to say that the World is immanent in God, as that God is immanent in the World.
> It is as true to say that God transcends the World, as that the World transcends God.

> It is as true to say that God creates the World, as that the World creates God.   (347–8)

That means that God Himself must be part of the evolutionary process. Not the world itself – pantheism – but in all parts of the world – what Whitehead, after Schelling, called "panentheism." And since the process of evolution involves struggle (the Englishman Whitehead is part of the Malthusian culture that produced Darwin), this means that God too must be suffering and striving: "he suffers from the suffering Inherent to the transitory." He has, as the story of Jesus shows full well, emptied Himself of His powers and works alongside us. "He made himself nothing by taking the very nature of a servant, being made in human likeness" (Philippians 2:7). This is known as kenosis. One should add, incidentally, that although he did not much like Whitehead's philosophy – his was a more conventional God – Dobzhansky was drawn to a Whiteheadian position. In his letters to Greene we read: "I see no escape from thinking that God acts not in fits of miraculous interventions, but in all significant and insignificant, spectacular and humdrum events. Panentheism, you may say? I do not think so, but if so then there is much truth in panentheism" (463).

Coming Full Circle

We have seen repeatedly, that a major – the major – difference between the mechanistic Darwinian evolutionist and the organicist non-Darwinian evolutionist is that, for the former, change comes from without – natural selection on random variations – whereas, for the latter, change comes from within – self-directing variations. In recent years, thanks to developments in molecular biology, there has been much interest in the forces taking mutations at the genetic level (genotype) to variations at the physical level (phenotype). Darwinian evolutionists are neither indifferent to nor inherently hostile to this new field – evolutionary development or "evo-devo" (see Arthur's superb book, *Understanding Evo-Devo*, in this series). They welcome it, but are nevertheless convinced that, as connections are unveiled, they will fit comfortably into a mechanistic framework. Organicist evolutionists such as Kevin Laland, to the contrary, seize on evo-devo as proof that they were right all along. Change will come from within as variations in some sense direct themselves.

Anti-Darwinian organicist, philosophers of mind Jerry Fodor and Massimo Piattelli-Palmarini are explicit in *What Darwin Got Wrong*: "the whole process of development, from the fertilized egg to the adult, modifies the phenotypic effects of genotypic changes, and thus 'filters' the genotypic options that ecological variables ever have a chance to select from" (27). It is not unfair to say that, often, there is a certain arrogance at work here, as we have seen in Medawar, with molecular biologists brushing aside whole-organism biologists, and thinking that they and they alone have the right strategy and answers. So, perhaps ignorant of the work of Dobzhansky, it is no surprise to find, right at the heart of his hostile review, Medawar writing:

> Unhappily Teilhard has no grasp of the real weakness of modern evolutionary theory, namely its lack of a complete theory of variation, of the origin of candidature for evolution. It is not enough to say that 'mutation' is ultimately the source of all genetical diversity, for that is merely to give the phenomenon a name: mutation is so defined. What we want, and are very slowly beginning to get, is a comprehensive theory of the forms in which new genetical information comes into being. It may, as I have hinted elsewhere, turn out to be of the nature of nucleic acids and the chromosomal apparatus that they tend spontaneously to proffer genetical variants – genetical solutions of the problem of remaining alive – which are more complex and more elaborate than the immediate occasion calls for . . .

Candidly, this could have come straight from the pen of that arch-*Naturphilosoph*, Friedrich Schelling.

4 Humans

For the Christian, if God is one end of the equation, we humans are the other end. What does this mean? What does this imply? The questions for this chapter.

Morality

Remember: "Let us make mankind in our image, in our likeness, so that they may rule over the fish in the sea and the birds in the sky, over the livestock and all the wild animals, and over all the creatures that move along the ground" (Genesis 1:26). It is there right at the beginning of the Old Testament; it is also in the New. We are unique and hence have a position superior to other animals. In the Sermon on the Mount, Jesus reaffirms this. "Behold the fowls of the air: for they sow not, neither do they reap, nor gather into barns; yet your heavenly Father feedeth them. Are ye not much better than they?" (Matthew 6:26). What does it mean to say that we are made in God's image, and that we are better than birds and, presumably, other animals also? In one way, obviously, the reference is to our senses and our reasoning powers. We need our senses and our reason to get on with life, to appreciate God's creation, and worship Him for it. "The heavens declare the glory of God; the skies proclaim the work of his hands" (Psalm 19:1). In another way, equally obviously, like God we are moral beings, we know the difference between right and wrong, and we know it is for us to follow the path of right rather than that of wrong. We know also that what we ought to do is very much not what we often do. We have wrong thoughts. We have wrong behaviors. Adam and Eve sinned, and they knew it. Fig leaves.

Now, the story starts to get a little more complicated. If we are made by an all-powerful, all-good God, then, surely, we must be good. Innately, we must always have the right thoughts and do the right thing. How to solve this paradox? One way is to deny or qualify some of the premises. We might question what it is to be "made" by God, if the manufacturing process must involve evolution, especially if it must involve (if Darwin is right) natural-selection-fueled evolution. Perhaps we will be led to question the assumption that God is all powerful. The process theologian will agree that in some sense God is all powerful, but part of that power – certainly part of that loving power – is that God restricts His power. He humbles himself. Kenosis. St Paul tells it like it is: "And being found in fashion as a man, he humbled himself, and became obedient unto death, even the death of the cross" (Philippians 2:8).

A God who humbles himself and shares the pain and suffering of humans is so much greater than a God who stands aloof – the God of Aquinas – unable by His very nature to enter fully into human emotions. For this kind of process-theology divinity, that humans are not perfect when they set out is the whole meaning of everything. It is for us to struggle, with God alongside us, to improve and move towards right thought and conduct; not to fail God and then expect God to put it right – as He did with the sacrifice on the Cross. Does this mean that after the fall (of Adam and Eve) we are inherently tainted, bad, and only the sacrifice can change us? Or that we are good but infused, as it were, with sin and the propensity so to sin? Let us leave this conundrum to the theologians. What they share is the assumption that humans have free will. It is for us to decide whether we are going to follow the path of the good or the path of the bad, of disobedience, of going against the will of God, however that may be spelt out.

Free Will

Does not that mean trouble right there? However you regard the evolutionary process, whether from the mechanist perspective or from the organicist perspective, you are supposing we are the product of law. Does this not mean we must follow the laws of nature and that consequently we can have no free will? In some sense, does it mean we are predetermined to do what we do and, just as you would not blame or praise a robot for what it does, so neither should you praise or blame humans for what they do? We know already, from the discussion in Chapter 2, that this cannot be the whole story. The point about humans, with their limited ability to give birth to vast numbers of offspring, let alone raise them to maturity, is that they must be able to respond to challenges. If it rains, mother ant does not altogether worry if a thousand of her offspring get washed away. There are lots more. If it rains, mother human worries a lot. Even if she has many children, like ten or twelve. "How many kids have you got?" "Ten. No, better qualify that. A couple of them went to McDonalds and it started to rain. Better say eight." For humans, if it starts to pour down, we can respond. Instead of going all the way to McDonalds: "Slip into Starbucks right here and have a latte." (Assuming, that is, that kids who go, of their own free will, to McDonalds would ever exercise that free will and go to Starbucks for a latte.)

Darwin's theory gives us understanding of why we need free will. For both mechanism and organicism, it is a challenge that must be addressed. Bergson wrestled with it. We know that he had a world picture that rejected mechanism. As an inhabitant of the land of Descartes, he had a somewhat indifferent attitude to animals. They may have a form of consciousness, but they do not have (mechanism-transcending) free will. Humans are different, says Bergson. Take two boys set the task of monitoring a primitive (Newcomen) steam-engine. The one stays on the job, turning taps on and off, as needed. The other ties up the handles with cords, so the machine turns itself on and off. This boy is now free to do other things as he will.

> . . . if we cast a glance at the two boys, we shall see that whilst one is wholly taken up by the watching, the other is free to go and play as he chooses, and that, from this point of view, the difference between the two machines is radical, the first holding the attention captive, the second setting it at liberty. A difference of the same kind, we think, would be found between the brain of an animal and the human brain. (*Creative Evolution*, 184)

Grant that the one boy is free in a sense that the other is not. Grant, what no Englishman would ever grant, that humans are free in a sense that dogs are not. We need still to know more about the nature of this freedom. There are two different solutions to the problem, according to Fischer, Kane, Pereboom, and Vargas, writing in 2007. The first position, "compatibilism," argues that not only is free will possible in a law-bound world, being law-bound is also a necessary condition for free will. If what you do is not governed by causes, then you have insanity not freedom. If they are doing it as part of a bet, the person who takes their clothes off in middle of campus during the day is free, even though you might think them very stupid to risk their studies for something so trivial. The person who starts to strip down for no reason whatsoever is simply a nutter. Interestingly, there are Christians – particularly Calvinists – who argue that compatibilism is mandated by their religion. God knows everything, including all about the future. In other words, we are predestined to do what we do. Without that, we impugn God's power. The great New England theologian Jonathan Edwards agrees: "The Will is always determined by the strongest motive." Motives don't come from nowhere. They come from philosophy professors teaching you to aspire to better things for yourself and others. If I did not believe that sincerely, and if I did not believe that this would have some causal effect on my students, I would never have taught for 55 years. Continuing this example, a professor hands out the exam to two students – the one has worked hard and studied, the other prefers the pleasures of the beer hall. The professor knows already who will do well and who will not, but that does not mean that the students had no free will.

Expectedly, coming from that dour land of Calvin's follower John Knox, north of Hadrian's Wall, David Hume was compatibilism's philosophical champion. Like Jonathan Edwards, he focused in on motives and argued that what happens is always a function of these motives, as what happens in the inanimate world is a function of prior causes.

> [It] appears, not only that the conjunction between motives and voluntary actions is as regular and uniform as that between the cause and effect in any part of nature; but also that this regular conjunction has been universally acknowledged among mankind, and has never been the subject of dispute, either in philosophy or common life. Now, as it is from past experience that we draw all inferences concerning the future, and as we conclude that objects will always be conjoined together which we find to have always been conjoined; it may seem superfluous to prove that this experienced uniformity in human actions is a source whence we draw inferences concerning them. (*Enquiry*, section 8)

For Hume, causes – and motives – don't exist "out there." They are part of our psychology that makes things understandable. And the only way we can understand human nature is by embedding it in a net of motives. Sometimes, like inexpensive rockets, once fired we cannot change direction in the face of obstacles. Sometimes, like the Mars Rover, we can alter directions according to the obstacles encountered. As Daniel Dennett says as subtitle to *Elbow Room*, his snappy little book on the topic, we have "The Varieties of Free Will Worth Wanting."

The second position on free will, "libertarianism" (not be confused with the political doctrine of libertarianism as promoted by Ayn Rand), argues that we do have free will and, in some sense, it is something outside of – beyond – the empirical *causus nexus*. The classic version of this is given in Kant's *Critique of Pure Reason*, where he argues that our powers of reason lift us above the blind working of unguided law.

> [T]he human will is not determined by that alone which immediately affects the senses; on the contrary, we have the power, by calling up the notion of what is useful or hurtful in a more distant relation, of overcoming the immediate impressions on our sensuous faculty of desire. But these considerations of what is desirable in relation to our whole state, that is, is in the end good and useful, are based entirely upon reason. This faculty, accordingly, enounces laws, which are imperative or objective laws of freedom and which tell us what ought to take place, thus distinguishing themselves from the laws of nature, which relate to that which does take place.

One suspects, presumes, that it is this notion of freedom that appeals to the organicist with the heavy emphasis – as we saw in Bergson – on the creative nature of the evolutionary process. There may be an upwards rise, but not without some creative, free input. The same would be true for the process philosopher or theologian. The inherently ongoing struggle to rise – that of humans and of God – is not something foretold or guaranteed. It is something all must work towards. Remember:

> To elicit from his creation its invention
> Of his own solace.
> (*Song of the World Becoming*,
> 182–3)

Substantive Ethics: Mechanism

Either way, allow that humans have some dimension of freedom – freedom of choice, that is. What are the basic questions to be asked about morality? An all-important distinction is between "what" questions and "why" questions. What ought I to do? Why should I do what I ought to do? Substantive (or normative) ethics, versus metaethics. Take Christianity as a paradigm, meaning (in a non-value sense) a system that speaks to both levels of morality. At the substantive level, the Love Commandment – "love your neighbor as yourself" – is at the heart of Christian morality. Leave elaborations until later. At the metaethical level, God's Will usually figures highly. "You ought to do what God wants you to do." This is Catholic natural law theory. The "light of natural reason, whereby we discern what is good and what is evil, which is the function of the natural law, is nothing else than an imprint on us of the Divine light. It is therefore evident that the natural law is nothing else than the rational creature's participation of the eternal law" (*Summa Theologica*, e I and II, 91, 2).

Start with substantive ethics. How does the mechanist speak to this? One must confess that there are times when Charles Darwin expresses sentiments (that have come to be known as "Social Darwinism") that do rather make one's hair stand on end.

> I could show fight on natural selection having done and doing more for the progress of civilisation than you seem inclined to admit. Remember what risks the nations of Europe ran, not so many centuries ago of being overwhelmed by the Turks, and how ridiculous such an idea now is. The more civilised so-called Caucasian races have beaten the Turkish hollow in the struggle for existence. Looking to the world at no very distant date, what an endless number of the lower races will have been eliminated by the higher civilised races throughout the world. (Letter 13230, to William Graham, July 3, 1881)

We have seen already how being a comfortably situated Victorian, at the height of the Empire's outreach, did influence Darwin's thinking. Note, though, that such comments as that just quoted tend to come in the privacy of letters, so perhaps Darwin himself realized they were more prejudice than carefully reasoned argument. In mitigation, note that the Darwin family, Charles very much included, were always strong against slavery. During the

American Civil War, in England, Darwin was one of a minority supporting the North. American slave-produced cotton was vital to the Lancashire fabric industry. Leaving weighing the merits or otherwise of Darwin's personal convictions, let us turn to public discussion of substantive ethics and its claims. It is in the *Descent of Man*, published in 1871, some 12 years after the *Origin*, that we find Darwin's thinking on the appearance and need of morality. He suggested that sociality could come through interaction of non-relatives, through what today is known as "reciprocal altruism" – you scratch my back, and I will scratch yours.

> In the first place, as the reasoning powers and foresight of the members became improved, each man would soon learn that if he aided his fellow-men, he would commonly receive aid in return. From this low motive he might acquire the habit of aiding his fellows; and the habit of performing benevolent actions certainly strengthens the feeling of sympathy which gives the first impulse to benevolent actions. Habits, moreover, followed during many generations probably tend to be inherited. (*Descent* 1, 163–4).

Then Darwin added that the praise and blame of our fellow humans would be a huge stimulus to the development of some form of moral sense. This is all bound up with sympathy, wanting praise and dreading criticism, and this is clearly something that came about through natural selection (1, 164). He elaborated: "To do good unto others – to do unto others as ye would they should do unto you, – is the foundation-stone of morality. It is, therefore, hardly possible to exaggerate the importance during rude times of the love of praise and the dread of blame" (1, 165).

He added:

> It must not be forgotten that although a high standard of morality gives but a slight or no advantage to each individual man and his children over the other men of the same tribe, yet that an advancement in the standard of morality and an increase in the number of well-endowed men will certainly give an immense advantage to one tribe over another. (1, 166)

There is no ambiguity about what this means. Tribes containing many members who pull their weight – patriots, faithful, obedient, courageous, sympathetic – willing to put their own interests on one side for the sake of all, would win out over other tribes. Natural selection! Hence, the consequence. "At all times throughout the world tribes have supplanted other tribes; and as morality is one element in their success, the standard of morality and the number of well-endowed men will thus everywhere tend to rise and increase" (1, 166).

Apparently, all very straightforward. Do not be deceived. There is here rather more than meets the eye. From the first, Darwin was always what today we would call a dedicated "individual selectionist." By this is meant that selection always works for the individual rather than the group. You might think that this could never be. Often groups compete and one group wins. Humans as they are today versus the Neanderthals. Group selection. Not quite so fast. If two groups compete, but it is the individual members that win or lose, this is still individual selection. It is only if the group wins as a group – presumably because some members of the group give to the whole without expectation of reward – that we can speak of "group selection." From the first, Darwin was against it. Possibly because he was from the upper-middle class, which had in the Industrial Revolution benefited from ruthless competition, he never thought that in nature we get the kind of disinterested altruism demanded by group selection. The co-discoverer of natural selection, Alfred Russel Wallace, was ever an ardent socialist, putting the group before the individual. He thought that hybrid sterility (as in the mule) is a function of the benefit to parent species (not wasting time on non-purebreds). Darwin (who thought the sterility of the hybrids a simple consequence of different reproductive systems not meshing) wrote:

> Let me first say that no man could have more earnestly wished for the success of N. selection in regard to sterility, than I did; & when I considered a general statement, (as in your last note) I always felt sure it could be worked out, but always failed in detail. The cause being as I believe, that natural selection cannot effect what is not good for the individual, including in this term a social community. (Letter to Wallace, April 6, 1868)

This is why, when discussing morality which is obviously (however caused) a group phenomenon, Darwin stressed reciprocal altruism. I help you, but very much on the expectation that you or others in the group will help me in return. Individual selection. One should note that Darwin is also appealing to another kind of process whereby we benefit from helping others. Relatives can benefit from your success. In the *Origin*, Darwin argued that if members of your family reproduce, then you, sharing elements of heredity, reproduce by proxy, as it were. "Thus, a well-flavoured vegetable is cooked, and the individual is destroyed; but the horticulturist sows seeds of the same stock, and confidently expects to get nearly the same variety; breeders of cattle wish the flesh and fat to be well marbled together; the animal has been slaughtered, but the breeder goes with confidence to the same family" (237–8).

This all rather supposes that Darwin thought that competing tribes were in fact competing groups of interrelated members. The supposition is correct. Following the comparative jurist Henry Maine, Darwin regarded tribes as interrelated families (or thinking they are), and he took the family to be one individual, a kind of super-organism. Without comment, showing his agreement, Darwin inserted (in the *Descent of Man*) a footnote: "After a time the members or tribes which are absorbed into another tribe assume, as Mr. Maine remarks (*Ancient Law*, 131), that they are the co-descendants of the same ancestors." Hence, with respect to morality, humans are like the ants. We are parts of a whole rather than individuals doing their own thing. Note, however, that although we may have a super-organism, the parts are furthering their own ends and only incidentally that of the whole. I am better off being part of a tribe.

Today's Darwinian evolutionists, who have taken up Darwin's thinking about reproducing by proxy – it is now called "kin selection" – have the same understanding of the evolution of morality. Hominins, human ancestors, broke from the apes five to seven million years ago, in Africa, leaving the jungle and moving out onto the plains. We became bipedal. No one is sure exactly why; but being able to stand and look for predators or prey is obviously a major adaptation, as is being able to keep going nigh indefinitely. As my cairn terriers show, four-legged beasts are faster, but we can simply keep after our prey until they drop from exhaustion, and then we can (literally) move in for the kill. We became hunter-gatherers on what one paleoanthropologist humorously referred to as our five-million-year camping trip. We lived in bands of about 50, and generally (given our paucity of numbers and the size of Africa) had little to do with other bands. Obviously, there was some interaction and transfer of members, but not much. Our brains grew accordingly from about 400 cubic centimeters to around 1,300 cubic centimeters, today's *Homo sapiens*, which appeared around half a million years go. Strictly speaking, we are the subspecies *H. s. sapiens*, for there were two other subspecies, the already-mentioned Neanderthals, *H. s. neanderthalensis*, and the Denisovans, *H. s. denisova*. The other two subspecies went extinct about 50 thousand years ago, for reasons still not completely understood. There was some interbreeding between subspecies. Folk in Europe have Neanderthal genes and folk in Asia, Denisovan genes; less so or not at all in Africa. The racist joke about being cavemen is on the Europeans, not the Africans.

What is clear is that the growth of brain size, connected obviously with intelligence, is a function of our sociability. Probably, a feedback process: the more social the more intelligence, the more intelligence the more social. This incidentally is why you should not worry too much about the Neanderthals having bigger brains than we do. If you think of brains as the hardware of computers, then, continuing the analogy, for the first 40 or 50 years of computers, the aim was to build bigger, stronger computers, able to do more difficult sums. But in the past 30 years or so, the aim has switched to being able to communicate – the worldwide web, email, and so forth. Humans generally and our subspecies particularly have entered that phase. Tool making for instance, and the ability to pass on abilities and training. Speech, most obviously. And more.

We had to have adaptations promoting and facilitating that social behavior, and obviously having a moral sense is a powerful step in that direction. You see a small child getting dangerously close to a river. Should you step in and help it, even though there may be a risk to you – especially if the child falls in? Morality urges us to help, even though there is a risk to us. Overall, the band is in better shape if we all think that way. Of course, there will be people who cheat. But other related adaptations are ready to jump in – for instance, a suspicion about people's actions, especially if they don't seem to jibe with what was said and done. Relatedly, anyone who has had children knows how innate is a sense of fairness: "It's not fair. He's got the biggest half." (Pointing out that it should be "bigger" does little to put things right. "It's still the biggest.") And so forth.

How does this all mesh with Christianity? Darwin seems to have thought that evolution points to Christian (substantive) morality. Remember: "To do good unto others – to do unto others as ye would they should do unto you, – is the foundation-stone of morality" (*Descent of Man* 1, 165). But is this the final word? Reciprocal altruism might stretch out to embrace the whole human species, but surely kin selection is more discerning. As Hume said: "A man naturally loves his children better than his nephews, his nephews better than his cousins, his cousins better than strangers, where everything else is equal. Hence arise our common measures of duty, in preferring the one to the other. Our sense of duty always follows the common and natural course of our passions" (*Treatise,* 483–4). Doesn't this go flatly against the ethic of the New Testament? After all, the Good Samaritan did not stop to inquire if the wounded stranger was in fact his first cousin. This kind of differential morality has no place in the teachings of Jesus.

A couple of points in response. First, many (most?) of us would defend a measure of differential morality. Not to do so, would itself be immoral. In Charles Dickens' *Bleak House,* the philanthropist Mrs Jellyby devotes all her time to the starving poor in Africa. She quite ignores the poverty and suffering on her own doorstep. The pathetic Jo, the crossing sweeper; not to mention the indifferent neglect of her own children. Most of us would agree with Dickens that there is something grossly immoral about Mrs Jellyby's thought and actions. Second, as always, the Bible is somewhat ecumenical on these issues. Countering the Good Samaritan is: "Anyone who does not provide for their relatives, and especially for their own household, has denied the faith and is worse than an unbeliever" (I Timothy 5:8). It seems that the Darwinian mechanist is not so very much out of line with Christianity.

Social Darwinism

Turn now to the organicist take on substantive ethics. If the Darwinian has troubles with Christian ethics, that seemingly is nothing to the troubles of the organicist. Start in with Herbert Spencer.

> We must call those spurious philanthropists, who, to prevent present misery, would entail greater misery upon future generations. All defenders of a Poor Law must, however, be classed among such. That rigorous necessity which, when allowed to act on them, becomes so sharp a spur to the lazy and so strong a bridle to the random, these paupers' friends would repeal, because of the wailing it here and there produces. (*Social Statics,* 323)

It is this kind of thinking that has given evolutionary ethics a bad name – a bad name because of the underlying struggle. Social Darwinism! Widows and children to the wall. In the words of Thrasymachus in Plato's *Republic,* "might is right." Anything further from Christianity it would be harder to imagine. "Jesus said, Suffer little children, and forbid them not, to come unto me: for of such is the kingdom of heaven. And he laid his hands on them, and departed thence" (Matthew 19: 14–15).

Before we dash on, however, there are a couple of points that should be made. First, the passage just quoted came early, perhaps even before Spencer became an evolutionist. It could be that the later sentiments, after he became an evolutionist – about cooperation and mutual aid – are more reliable. Second, Spencer is not really suggesting that widows and children should go the wall, end of argument. As someone lower-middle class, in Victorian England, he was only too aware that the deck was stacked in favor of the haves over the have-nots. If you were rich, with status and education, life could be sweet. Not otherwise. What he was really advocating were opportunities for the bright, the hardworking, the ambitious. No barriers for talent and industry. In every respect, he was a forerunner of Margaret Thatcher, like Spencer from a lower-middle class, non-conformist (Protestant, non-Anglican), Midlands background. She too wanted to remove barriers preventing the talented from moving upwards. Indicative was the fight over the grammar schools, the schools for those children who could pass the 11-plus exam – a combination test of IQ and the three Rs. Many, including most socialists, wanted to abolish them as elitist. Thatcher, with a surprising cross-political party support, practically repeated Spencer's words. Right or wrong, this was not Social Darwinism as it is usually understood.

Spencerian-type thinking transferred readily to the United States, particularly to those northern cities now in the throes of an industrial revolution that was, by the twentieth century, to move the country to world forefront. Well known is the Scottish-born Andrew Carnegie, who not only built a massive steel works (later bought out and amalgamated as part of US Steel), but who also showed ruthless determination in breaking workers in the notorious Homestead Strike of 1892, replacing them with non-union immigrants. He justified his behavior with neo-Spencerian, Social Darwinian sentiments.

> The price which society pays for the law of competition, like the price it pays for cheap comforts and luxuries, is also great; but the advantages of this law are also greater still, for it is to this law that we owe our wonderful material development, which brings improved conditions in its train. But,

whether the law be benign or not, we must say of it, as we say of the change in the conditions of men to which we have referred: It is here; we cannot evade it; no substitutes for it have been found; and while the law may be sometimes hard for the individual, it is best for the race, because it insures the survival of the fittest in every department. (*Gospel of Wealth*, 655)

Yet, as with Spencer, the story is more complicated for, as is very well known, having made his fortune, Carnegie proceeded to give it away! In his *Gospel of Wealth*, he asked about what one should do with all this wealth once one had accumulated it. He saw three responses.

First, you do the traditional thing and leave it to your family. This is "most injudicious"! "Beyond providing for the wife and daughters moderate sources of income, and very moderate allowances indeed, if any, for the sons, men may well hesitate, for it is no longer questionable that great sums bequeathed oftener work more for the injury than for the good of the recipients." Carnegie knew all about idle playboys.

The second is to leave your money at death for the furtherance of good works. Bad idea! "Knowledge of the results of legacies bequeathed is not calculated to inspire the brightest hopes of much posthumous good being accomplished. The cases are not few in which the real object sought by the testator is not attained, nor are they few in which his real wishes are thwarted. In many cases the bequests are so used as to become only monuments of his folly."

This leaves the third option: use your money yourself for the good of society. "Under its sway we shall have an ideal state, in which the surplus wealth of the few will become, in the best sense the property of the many, because administered for the common good, and this wealth, passing through the hands of the few, can be made a much more potent force for the elevation of our race than if it had been distributed in small sums to the people themselves." Carnegie is justly famous for his saying that "no man should die rich." He is as famous for putting his beliefs into practice and sponsoring the founding of public libraries. I am sure the childhoods of many readers, as was mine, were enriched by Carnegie's generosity.

Despite the qualifications and excuses one can make, the meaning of the label "Social Darwinism," meaning something harsh and unforgiving, stuck; and through the twentieth century and down into the twenty-first it is used as a pejorative term only. Nevertheless, whatever it was called – or not called – a more constructive and meaningful organicist approach to substantive ethics persisted. We have seen Julian Huxley's organicist inclinations, his enthusiasm for Bergson and (later) Teilhard de Chardin. His epistemological convictions were equalled by his ethical enthusiasms. As a scientist, writing during the 1930s and 1940s, Julian Huxley thought – not surprisingly – that (particularly at the societal level) we should be promoting the virtues and benefits of science and technology. Responding to the Great Depression, we find that Huxley was a great enthusiast for the public works funded by Franklin Roosevelt's New Deal. Although stepping somewhat warily because he did not want to be seen as endorsing the war preparations of the National Socialists – the building of the Autobahn for example – Huxley was unrestrained in his encomia for the Tennessee Valley Authority, that project bringing electricity to large parts of the American South. After the Second World War, Huxley became the first Director-General of UNESCO. It was he who insisted that the United Nations go beyond just education and culture to include science also.

The mechanist worked from an individual selection perspective. Does the organicist work from a group selection perspective? It is not always easy to answer this question. Spencer does not rely much on selection at all in his theorizing. However, it is true that if you are searching for group selection, organicism is a promising field. The recently deceased, major biologist Edward O. Wilson was an ardent Spencerian. Amusingly, but also revealingly, he had a larger picture of Spencer on his wall than that of Darwin. He shared Spencer's holism, to the extent that (much to the horror of his fellow evolutionists) he was an ardent defender of group selection: "four decades of research since the 1960s have provided ample empirical evidence for group selection, in addition to its theoretical plausibility as a significant evolutionary force" (*Rethinking*, 334). In line with this: "[in] virtually all cases, traits labeled cooperative and altruistic are selectively disadvantageous within the groups and require between-group selection to evolve" (335). Extending discussion to our area of interest, he goes on: "Group selection is an important force in human evolution in part because cultural processes have a way of creating phenotypic variation among groups, even when they are composed of large numbers of unrelated individuals" (343). To parody Rabbi Hillel (who, when asked to explain the Torah in the time that he could stand on one foot, famously replied: "Do not do unto others that which is repugnant to you. Everything else is commentary"), in the world of biology: "Selfishness beats altruism within groups. Altruistic groups beat selfish groups. Everything else is commentary" (345).

Australian philosopher Peter Singer argued in 1972: "The fact that a person is physically near to us, so that we have personal contact with him, may make it more likely that we shall assist him, but this does not show that we ought to help him rather than another who happens to be further away" (*Famine*, 232). Singer is not arguing from a Christian perspective but rather from his version of utilitarianism – a version which has led Singer to argue for the moral need to kill severely handicapped newborns. Unsurprisingly, his thinking is anathema to many Christians. This said, his claim about the universal scope of morality is one that, as we have seen, reflects the Parable of the Good Samaritan. Also, it points to the disconnect between what many think is the organicist position on (substantive) morality – Social Darwinism – and the true organicist position on such morality. If organicists at times embrace a doctrine of nature red in tooth and claw, it is generally in a limited context and only ancillary to their essential belief in the organic unity of society and the need to improve the wellbeing of all its members. Whether this version of Christian ethics – morality applies indifferently across the whole group – is more plausible and supported than the alternative version – charity starts at home – we can leave to church discussion groups and move on.

Debunking

We are being pointed to questions of metaethics. The modern-day Darwinian agrees with Darwin. Morality is a good adaptation. Does this mean that morality is, as the logical positives said, nothing more than emotions? I don't like cruelty. Boohoo, don't you be cruel. Clearly this is true in a sense, although morality must be a special kind of emotion, where we – to use a rather ugly term of John Mackie – "objectivize" our emotions. I think you should help young people, not just because I don't like it when you don't help, but because I think you have an objective moral reason to help young people.

This is not our choice. It is what is really right or wrong. That is why I have the authority to judge you and to order you to obey. What we think right or wrong is very much a function of our evolution.

> I do not wish to maintain that any strictly social animal, if its intellectual faculties were to become as active and as highly developed as in man, would acquire exactly the same moral sense as ours. In the same manner as various animals have some sense of beauty, though they admire widely different objects, so they might have a sense of right and wrong, though led by it to follow widely different lines of conduct. If, for instance, to take an extreme case, men were reared under precisely the same conditions as hive-bees, there can hardly be a doubt that our unmarried females would, like the worker-bees, think it a sacred duty to kill their brothers, and mothers would strive to kill their fertile daughters; and no one would think of interfering. Nevertheless the bee, or any other social animal, would in our supposed case gain, as it appears to me, some feeling of right and wrong, or a conscience. (*Descent of Man* 1, 73)

We are humans not hive-bees and, thank God or Darwin, we males do not have to worry as winter approaches. But the relativity point is made. Objective standards are shown to be impossible. To use a popular term among philosophers, they have been "debunked."

"If it's what you want, then it's okay." Isn't this echo from the 1960s deeply antithetical to the Christian take on foundations? This latter is not relative. Morality in some sense reflects the will of God. The trouble with this statement unadorned is that, as Plato noted in his dialogue the *Euthyphro*, one is left dangling between two somewhat unacceptable alternatives. If everything is the will of God, who is or was there to control Him? Surely God could not make it morally acceptable to beat little old ladies on the head, just for fun? But if God is rather reflecting eternal objective values, then His omnipotence seems circumscribed. He is not the ultimate source of morality. The solution to this conundrum is Catholic natural law theory. We are to do what God dictates, but what He dictates is a function of how things have been created. What is "natural"? God has given us two sexes, so heterosexual intercourse is natural and hence (within constraints like not having sex with children) morally acceptable. Bestiality, sex with other animals, is not natural and hence immoral. Had God made us, like some Amazonian fish, needing members of another species to spur their asexual reproduction, then bestiality would be natural and hence moral.

All of this is music to the ears of the Darwinian. Doing good is doing what is natural. End of argument.

Progress

What of the organicist take on metaethics? The answer is one word: progress. Morally, we ought to do that which aids progress – ultimately, progress to human beings. Spencer is categorical: from the simple to the complex, from the homogeneous to the heterogeneous. Remember: "this law of organic progress is the law of all progress." No matter if you are talking about the Earth itself, life on the Earth, or human culture: "this same evolution of the simple into the complex, through successive differentiations, holds through-out" (*Progress*, 245). Hence morality emerges through the evolutionary process, and our duties are to ensure that this happens by removing barriers and facilitating the process. "Ethics has for its subject-matter, that form which universal conduct assumes during the last stages of its evolution" (*Data of Ethics*, 21). Spencer continues: "And there has followed the corollary that conduct gains ethical sanction in proportion as the activities, becoming less and less militant and more and more industrial, are such as do not necessitate mutual injury or hindrance, but consist with, and are furthered by, co-operation and mutual aid" (20).

Thomas Henry Huxley would have none of this. For him, morality came in fighting our animal nature. "For his successful progress, throughout the savage state, man has been largely indebted to those qualities which he shares with the ape and the tiger; his exceptional physical organization; his cunning, his sociability, his curiosity, and his imitativeness; his ruthless and ferocious destructiveness when his anger is roused by opposition." Nevertheless, "in proportion as men have passed from anarchy to social organization, and in proportion as civilization has grown in worth, these deeply ingrained service-able qualities have become defects. After the manner of successful persons, civilized man would gladly kick down the ladder by which he has climbed. He would be only too pleased to see 'the ape and tiger die'" (*Evolution and Ethics*,

52). Yet, although grandfather Huxley had little time for Spencerian-type thinking, his biologist grandson Julian Huxley was an enthusiast. He was an ardent progressionist, thinking that as we go up the scale, value increases – things get better. "When we look at evolution as a whole, we find, among the many directions which it has taken, one which is characterized by introducing the evolving world-stuff to progressively higher levels of organization and so to new possibilities of being, action, and experience" (*Evolutionary Ethics*, 41–2). Spelling things out, he says: "Increase of control, increase of independence, increase of internal co-ordination; increase of knowledge, of means for co-ordinating knowledge, of elaborateness and intensity of feeling—those are trends of the most general order. If we do not continue them in the future, we cannot hope that we are in the main line of evolutionary progress any more than could a sea-urchin or a tapeworm." One hardly need add: "The future of progressive evolution is the future of man." Huxley incidentally put his money where his mouth was. To mark his appointment as director general of UNESCO, he wrote a little book, *UNESCO: Its Purpose and Its Philosophy*, praising progress and declaring it the philosophy of the new organization. His overseers were so shocked, they cut his intended term from four years to two.

"Ought" from "Is"?

From the beginning, from before, philosophers have been highly critical of this line of argument. The objection goes back to Hume. We should not slide without notice from talking about what is, to what ought to be. "For as this ought, or ought not, expresses some new relation or affirmation, 'tis necessary that it should be observed and explained; and at the same time that a reason should be given, for what seems altogether inconceivable, how this new relation can be a deduction from others, which are entirely different from it" (*Treatise*, 302). Post-Darwinian philosophers made this their trademark objection to evolutionary ethics. Henry Sidgwick, in the first year of the journal *Mind*, stated bluntly that the theory of evolution, as was understood at the time, has little or no bearing upon ethics. Spencer is the focus of attack, as he is in the much-celebrated book, *Principia Ethica*, by Sidgwick's student G. E. Moore.

He derided the mistaken attempt to go from claims about matters of fact to claims about matters of obligation, labeling it the "naturalistic fallacy." Herbert Spencer is identified as a grave offender. Hypotheses about evolution tell us what is more evolved than others. They tell us nothing about ethical merit (31). Moore continues that, although it may be that what is "more evolved" is also higher and better, Spencer does not show this in any way. Indeed, he seems to be unaware that there is any difference.

Although non-organicists may find this line of argument convincing, it has little effect on organicists. For someone like Spencer – Julian Huxley likewise – the living world is value-impregnated. We are not going from "is" to "ought." We are going from "ought" to "ought." We are looking at an organism in growth, and as the oak tree is superior to the acorn, the butterfly to the caterpillar, so, as we go from monad to man, value increases and our moral obligations – to aid this process – come from them. Edward O. Wilson was a paradigm of this way of thinking. He was deeply committed to organic progress: "Four groups occupy pinnacles high above the others: the colonial invertebrates, the social insects, the nonhuman mammals, and man" (*Sociobiology*, 379). He continues: "Human beings remain essentially vertebrate in their social structure. But they have carried it to a level of complexity so high as to constitute a distinct, fourth pinnacle of social evolution" (380). For him, therefore, no argument was needed to conclude that this all justifies morality.

> Camus said that the only serious philosophical question is suicide. That is wrong even in the strict sense intended. The biologist, who is concerned with questions of physiology and evolutionary history, realizes that self knowledge is constrained and shaped by the emotional control centers in the hypothalamus and limbic systems of the brain. These centers flood our consciousness with all the emotions – hate, love, guilt, fear, and others – that are consulted by ethical philosophers who wish to intuit the standards of good and evil. What, we are then compelled to ask, made the hypothalamus and limbic system? They evolved by natural selection. That simple biological statement must be pursued to explain ethics and ethical philosophers, if not epistemology and epistemologists, at all depths. (*Sociobiology*, 3)

Where does religion figure in all of this? None of the organicists we have mentioned were writing explicitly from a Christian perspective. Herbert Spencer was an agnostic. Andrew Carnegie hated Christianity:

> The whole scheme of Christian Salvation is diabolical as revealed by the creeds. An angry God, imagine such a creator of the universe. Angry at what he knew was coming and was himself responsible for. Then he sets himself about to beget a son, in order that the child should beg him to forgive the Sinner. This however he cannot or will not do. He must punish somebody – so the son offers himself up & our creator punishes the innocent youth, never heard of before – for the guilty and became reconciled to us. ... I decline to accept Salvation from such a fiend. (Letter from Carnegie to Sir James Donaldson of St Andrews University, 1905)

Although not as vehemently anti-religion, Julian Huxley was explicitly atheistic. And Wilson had moved on from his Evangelical childhood. That said, leaving aside the already-discussed issue of how readily an organicist can eschew all belief in a deity, much if not all that they said would – should – find the approval of the Christian. For people from Spencer's background – Margaret Thatcher's too – the Parable of the Talents always loomed large in their theology. God did not put us on this world just to have a good time. We are expected to work and to use the gifts that we have. Making this possible by removing social barriers or, as in the case of Carnegie, offering help along the way is doing the Lord's work. Again, the Tennessee Valley Authority, making electricity available across the South, is a paradigmatic instance of following the Love Commandment, as is directing UNESCO, especially seeing that the benefits of science can become more freely available. Wilson, as we shall see later, likewise fell into this camp. He took pride in being more Christian than the Christians! And that is a good note on which to end this chapter.

5 Environment

Environmental Challenges

Let us return to the address to the American Association for the Advancement of Science given in 1967 (later published in *Science*), by medieval historian Lynn White Jr. He threw down the gauntlet. The son of a Presbyterian minister and himself a lifelong active Presbyterian, White felt nevertheless that his religion had given rise to much that needed answering. "Christianity bears a huge burden of guilt." White's thesis is relatively straightforward. Modern science and technology – and the appalling environmental consequences – are the children of the Christian faith. But we get ahead of ourselves. Let's plunge right into looking at some of these environmental consequences, using as our exemplar the most pressing environmental issue of them all: global warming.

It has been recognized since the eighteenth century that the Earth has been subject to heating and cooling; the latter times being known as "ice ages." The evidence for times of cooling is varied. Particularly powerful is the existence of such phenomena as "erratic boulders," that is, chunks of rock of one kind found in places where the general underlying rock is of another kind. Those living in countries with glaciers, such as Switzerland, recognized that the reason for erratic boulders is the action of glaciers (formed during cold periods) that carry them from one site to another, depositing them as the temperature warms up and the ice melts. One of the most famous instances of such an inference occurred in the late 1830s in Scotland. The young, ambitious Charles Darwin argued that the "parallel roads" of Glen Roy – seeming paths going in parallel along the sides of the valley (see

Figure 5.1   The parallel roads of Glen Roy.

Fig. 5.1) – were the shores of the sea, which had once filled the glen, and which now had retreated owing to the rising level of the surrounding land (a result of the general waterbed-like up and down movements of the land, as postulated by Charles Lyell's uniformitarian theory of climate). The Swiss scientist Louis Agassiz, visiting the glen, reasoned correctly that the paths were the shores of a lake, once contained in the glen by a glacier that, from the evidence of erratic rocks and scouring of the land, had now melted releasing the waters as the climate warmed. Darwin's big mistake! (He did get a publication out of it in the prestigious *Philosophical Transactions of the Royal Society of London*.)

The reason for ice ages was long debated, but today it is generally agreed that they are the result of continental drift. Thanks to plate tectonics, the continents move around the globe, and, as they do, they alter water patterns, and these in turn influence climate. London is at the same latitude as the city of Calgary in Alberta, Canada, but whereas the former has a temperate climate, the latter is very cold. London is a lot warmer simply because the Gulf Stream, coming up across the Atlantic from the Caribbean, heats the British Isles to a temperature much higher than one would expect if given only the information about latitude. Note that the theory of plate tectonics is a direct descendent of the eighteenth-century geological theory of Vulcanism of the Scottish farmer James Hutton, who argued that the center of our planet is molten rock, which occasionally breaks through the surface where it solidifies and moves across the globe as it is pushed by new effusions. Eventually the solid rock finds itself without room and is forced back down into the planet, where it melts and the whole process starts again.

Note also that Vulcanism – and in its turn plate tectonics – is a reflection of the machine model of modern science. In the eighteenth century, mining grew increasingly important – coal, metals, and so forth. Ever deeper mines called for ever more sophisticated ways of pumping water up and away. Hence, the Newcomen engine. It worked by producing steam, which then goes in a cycle of condensation and reheating, as it drives a pump (Fig. 5.2). It would be harder to think of a closer humanmade counterpart to the circular mechanism driving and creating the Earth in the Vulcanist picture of things. The sixteenth-century machine metaphor has conquered over the Platonic organic metaphor. And that spells an end to talk of final causes and goals and such things. Machines considered as things we use have purposes. The Newcomen engine is for pumping water, which we want it to do. Machines considered as machines – as in science – are just matter in motion, governed by eternal, unchanging laws, and have no purpose. Hence, there is no purpose to the Newcomen-like cycling globe. It works, and that is all you can or need to say.

Gaia

Organicists can hardly deny any of this, as such. There is, however, somewhat of a tendency to downplay the significance of ice ages and the fluctuations that they signify. The emphasis is always much more on the holistic side of things. This means directly that the discussions rarely if ever are purely about geology and physical climatology and so forth – the kind of discussion that we have just been having in the paragraphs above. The organic is as much an integral part of the discussion as the inorganic.

Expectedly, Alfred North Whitehead and his followers were both cause and effect of this movement. The organismic view of nature was the very essence of Whitehead's philosophy. In *Science and the Modern World*, Whitehead wrote:

> The trees in a Brazilian forest depend upon the association of various species of organisms, each of which is mutually dependent on the other species. A single tree by itself is dependent upon all the adverse chances of shifting circumstances. The wind stunts it: the variations in temperature

Figure 5.2   Newcomen engine.

check its foliage: the rains denude its soil: its leaves are blown away and are lost for the purpose of fertilisation. You may obtain individual specimens of fine trees either in exceptional circumstances, or where human cultivation has intervened. But in nature the normal way in which trees flourish is by their association in a forest. Each tree may lose something of its individual perfection of growth, but they mutually assist each other in preserving the conditions for survival. The soil is preserved and shaded; and the microbes necessary for its fertility are neither scorched, nor frozen, nor washed away. A forest is the triumph of the organisation of mutually dependent species. (206)

This kind of thinking leads naturally to the idea of the world as an organism – what James Lovelock has called the "Gaia" hypothesis. Its very essence is holism/organicism. Lovelock was by training a chemist, and unsurprisingly he thought as a chemist. For him, what really counted for something to be an organism was homeostasis – a body being able to keep itself in balance. Not a silly thought when you consider. Humans keep a constant body temperature through sweating and shivering. A lump of rock does not. He pointed out that, considering heat from the Sun and the cooling of the Earth through heat being lost, we should first have a cooling of the Earth since its formation and then a rise. What we have in fact is an almost stable temperature (Fig. 5.3), suggesting that somehow the Earth is in balance. Homeostasis! What better proof could one have that the Earth is an organism?

Marshaling the evidence, it is true that sometimes Lovelock thinks as a mechanist. One powerful item in favor of homeostasis is the fact that the salinity of the sea has apparently been stable, whereas one might have expected it to get more and more salty, like the Dead Sea. To explain this, Lovelock worked entirely within the framework of normal, mechanistic processes, fueled by efficient causes, working in a kind of feedback mode. Hypothesizing that the salt level of the sea is linked to the amount of silica in the water, Lovelock wrote of one such process for keeping things stable:

> This biological process for the use and disposal of silica can be seen as an efficient mechanism for controlling its level in the sea. If, for example, increasing amounts of silica were being washed into the sea from the rivers, the diatom population would expand (provided that sufficient

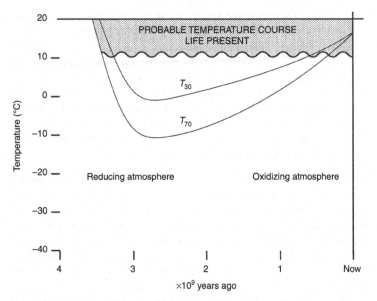

Figure 5.3   Gaia homeostasis. T_{30} is the Earth's surface temperature predicted with a 30 percent increase in solar luminosity (radiant energy from the sun); T_{70} with solar luminosity increase at 70 percent.

nitrate and sulfate nutrients were also in good supply) and reduce the dissolved silica level. If this level fell below normal requirements, the diatom population would contract until the silica content of the surface waters had built up again, and this is well known to occur. (*Gaia*, 88–90)

No cause for comment here. Just what one would expect from someone working under the machine metaphor. But then at once Lovelock showed that he, *as a scientist*, was putting this in a teleological context. "This deluge of dead organisms [diatoms] is not so much a funeral procession as a conveyor belt constructed by Gaia to convey parts from the construction zone at surface levels to the storage regions below the seas and continents." To spell it out: "**constructed** by Gaia **[in order] to** convey parts" (*Gaia*, 90, my emphasis). Organicist to the core.

Global Warming

With this background, turn now to Global Warming. Already we are primed. Mechanists are going to think of it in their terms, meaning it was caused by machine-like factors and must (and can) be cured by machine-like factors. Organicists, especially those who think in a Gaia perspective, are going to see something abnormal, something threatening homeostasis. The causes are in some sense out of the normal ordering of things, and cures, if possible, will require something similar. What then is Global Warming and why is it occurring? Beyond all reasonable dispute, the facts are these. First, there is an increase in average global temperature. As noted above, there have always been fluctuations in the general temperature of the globe, or at least parts of the globe. Ice ages are well documented. Apparently, today, we are in the reverse of an ice age. Over the past century and a half, there has been a significant rise in the average global temperature. Simultaneously, the sea level has risen and the snow cover of the Northern Hemisphere – especially up close to the Arctic – has decreased. It is estimated that between 1880 and 2012 the average surface temperature of the Earth has increased by about one degree Celsius; more, if you put the starting date back to the eighteenth century and the beginning of the Industrial Revolution (Fig. 5.4).

What is beyond doubt is that this rise in temperature is not natural, in the sense of being caused by non-human forces. As we have seen, it is suspected that, thanks to plate tectonics, natural climate changes might follow on the shifting of the land–sea balance. All agree that, in the case of the recent rise (that which we are experiencing), it is brought on by human activities, and that if we continue on course, we can expect a speeding up of temperature rise, perhaps by two degrees or more by the end of this century. Expectedly, there is debate about the exact causes and their relative importance, but general agreement is that very significant is the ever greater reliance on fossil fuels. Why is this? If it is to be habitable at all, the Earth needs its atmosphere containing the "greenhouse gases" which prevent the heat from the sun dissipating almost immediately. As in a greenhouse, solar heat is trapped, and the general temperature of the Earth rises, enough to bear life. The chief greenhouse gas is carbon dioxide, produced naturally by animals and by geological phenomena such as volcanoes. The use of fossil fuels produces carbon dioxide artificially and, with the ravaging of forests (like the Brazilian jungle), the trees of which absorb carbon dioxide via photosynthesis, unrestrained the gas enters the atmosphere and traps energy from the sun, with consequent temperature rise.

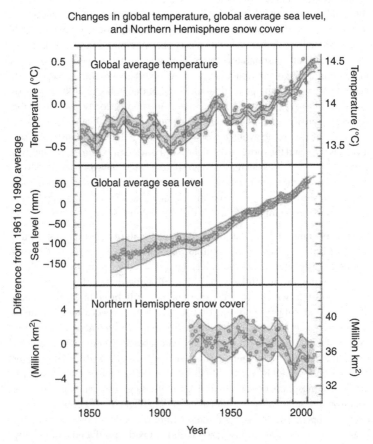

Figure 5.4   Global Warming. (All changes are relative to corresponding averages for 1961–1990.)

Coal, as we shall learn, has a long history of causing pollution. Nevertheless, it is not chance that such fuels started to be used significantly just at the time when the machine metaphor was pushing the organic metaphor to one side. Most immediately and obviously, as pointed out by the mineralogist Georgius Agricola, the organic metaphor seemed to block progress, particularly progress that involved the invention and use of aids to labor, machines. First, Agricola gave the traditional argument, based on the organic metaphor, that the move to machines demanded unacceptable practices.

> The earth does not conceal and remove from our eyes those things which are useful and necessary to mankind, but on the contrary, like a beneficent and kindly mother she yields in large abundance from her bounty and brings into the light of day the herbs, vegetables, grains, and fruits, and the trees. The minerals on the other hand she buries far beneath in the depth of the ground; therefore, they should not be sought. But they are dug out by wicked men who, as the poets say, "are the products of the Iron Age." (*De Re Metallica*, 6–7; the translation is by the future president of the US, Herbert Hoover, who was a mining engineer at the time)

Then, on pragmatic grounds, he went after the argument:

> If we remove metals from the service of man, all methods of protecting and sustaining health and more carefully preserving the course of life are done away with. If there were no metals, men would pass a horrible and wretched existence in the midst of wild beasts; they would return to the acorns and fruits and berries of the forest. They would feed upon the herbs and roots which they plucked up with their nails. They would dig out caves in which to lie down at night, and by day they would rove in the woods and plains at random like beasts, and inasmuch as this condition is utterly unworthy of humanity, with its splendid and glorious natural endowment, will anyone be so foolish or obstinate as not to allow that metals are necessary for food and clothing and that they tend to preserve life? (14)

Minerals and increasingly the fuel that made it all possible – coal.

Two Paradigms Revisited

It is nigh tautological that Global Warming has elicited responses that again show the divide between mechanistic and organismic thinking. On the one side, the immediate response is not to give up and try to return to the eighteenth century, but to see how mechanistic science can alleviate the problems. One obvious way is to turn to alternative sources of energy, reducing our reliance on fossil fuels. This is already happening, if in a limited way. For instance, in America and a number of other countries the use of coal is declining dramatically. True, much of this is due to turns to other fossil fuels, natural gas for instance. More controversial has been the use of nuclear methods of producing electricity. Three Mile Island (1979, USA), Chernobyl (1986, Soviet Union), and Fukushima (2011, Japan) come to mind. Less tense-making, wind and solar power are starting to make a real difference; in the United Kingdom in 2020, about 40% of the electricity produced came from renewable production. In the 20 years that he has lived in Florida, the author has moved entirely from relying on coal-produced electricity to solar-produced energy.

On the other side, as expected, there have been calls for holistic approaches. Move to North Dakota or somewhere likewise unappealingly beyond civilization, build your own home deliberately eschewing such unneeded luxuries as indoor plumbing, eat vegan (of course), and remember no self-respecting organicist can take a step beyond the front door unless they are wearing Birkenstocks. Compensating, this philosophy usually means the obligation to sleep with someone else's wife (or husband; or both). As Jan Smuts put it in 1926, "Holism is a process of creative synthesis, the resulting wholes are not static but dynamic, evolutionary, creative" (89). Can't say fairer than that. It is true that there is a little bit of this attitude in the mechanist's approach. Apart from the attractions of extra-marital sex, as noted above, tearing up the Brazilian rainforests increases carbon dioxide in the atmosphere, which spells more heat. The organicist starts with this kind of thing and goes from there. For the organicist, it is not just enough to start searching for solutions. A kind of spiritual awakening is demanded too. If we follow Fritjof Capra in speaking of mechanism as "the view of the universe as a mechanical system composed of elementary building blocks, the view of the human body as a machine, the view of life in society as a competitive struggle for existence, the belief in unlimited material progress to be achieved through economic and technological growth, and last but not least, the belief that a society in which the female is everywhere subsumed under the male is one that follows a basic law of nature," it is sadly concluded that it will not do. We need a "fundamental change of world view" (19–20). "Competitive struggle," "unlimited material progress," "female everywhere subsumed under the male." Need we say more? The world is a harmonious entity with its own intrinsic worth.

The Green Revolution

As we turn to religious implications, it is worth noting in passing that the philosophical divide that we see in the Global Warming debate repeats itself again and again when environmental issues are raised. A good example is the Green Revolution, a movement from around 1950 to around 1970, aimed at improving agricultural yields dramatically, to speak to widespread poverty and hunger, particularly in the Developing World. On the one side, it was as firmly entrenched in the machine paradigm as it was possible to be. It had several related strategies, but above all the creation of high-yielding varieties of cereals, especially dwarf wheat and rice. More recently, there has been the "genetic modification" of organisms (GM foods), as genes were artificially added or removed from natural varieties. Famous, notorious, is Golden Rice. This is a variety of rice (*Oryza sativa*) modified to produce, through genetic engineering, in the edible parts of rice, more beta-carotene (a strongly colored red–orange pigment abundant in fungi). Why is this important? It is a precursor of vitamin A. Given that rice is the staple crop for over half of the world's population and given that this new kind of rice can prevent vitamin A deficiency, its potential for good needs no stress. Or if it does, let it be mentioned that the lack of vitamin A is thought to cause nearly 700 thousand early childhood deaths and another 500 thousand cases of irreversible blindness (annually).

Expectedly, this has not met with universal approval, and as expectedly the pollution of the environment has figured large in the criticisms. Ed Regis writes that non-stop planting of a few genetically related high-yielding varieties, often jammed right in next to each other, has led to the appearance of new kinds of insect pests. Heavy fertilizer applications are producing nitrate levels in drinking water that exceed tolerable levels, while pesticides have eliminated fish and weedy green vegetables from the fields and thus in the diet of poor farmers. As expectedly, holistic/organismic issues ride high. Indian activist Vandana Shiva writes of Golden Rice that its use "is a recipe for creating hunger and malnutrition, not solving it." Of the claim that it can significantly reduce problems of blindness affecting children in the Developing World, she writes: "it appears as if the world's top scientists suffer a more severe form of blindness than children in poor countries." We learn that claims that "traditional breeding has been unsuccessful in producing crops high in vitamin A" are not true "given the diversity of plants and crops that Third World farmers, especially women, have bred and used which are rich sources of vitamin A such as coriander, amaranth, carrot, pumpkin, mango, jackfruit." Note that driving Shiva as much as the technological questions is her philosophy – that we need to return to traditional methods, more in tune with custom and nature. Explicitly, Shiva acknowledges that she approaches these questions from a holistic perspective. As always, Darwin is to blame: "the symbioses, the interconnections that nurture and sustain life are ignored, and both natural evolution and social dynamics are perceived as impelled by a constant struggle of the stronger against the weaker, of constant warfare." We need a holistic perspective. On which note, let us turn back to Lynn White's excoriation of Christianity's role in all of this.

Christianity – Historical Questions

Christianity privileges humans above all others: we are made in "God's image," says White (*Historical Roots*, 1205). As a consequence, Christianity set humans apart from the rest of nature, and insisted that God intended humans to use nature for their own legitimate ends (1205). Naturally, honoring our God, we use our reason and set about living in the world He has created – He has created for us. And this has had predictable consequences. For example, with the coming of the automobile, there has been a huge decline in the sparrow population. No horse manure to pick through and eat. Historically, even by 1285, London had a smog problem arising from the burning of soft coal. This was minimal by today's horrendous standards: "With the population explosion, the carcinoma of plan-less urbanism, the now geological deposits of sewage and garbage, surely no creature other than man has ever managed to foul its nest in such short order" (1204). Repent! Repent! In conclusion, White says: "we shall continue to have a worsening ecologic crisis until we reject the Christian axiom that nature has no reason for existence save to serve man" (1207).

Needless to say, many rushed to pick up lances and to ride out in defense of Christianity. Unsurprisingly, much criticism was directed against White's reading of the Bible. It was pointed out that there are lots of passages that stress concern for animals and so forth. "Woe to the worthless shepherd, who deserts the flock! May the sword strike his arm and his right eye! May his arm be completely withered, his right eye totally blinded!" (Zechariah 11:17). The cynical will point out that the shepherd is hardly looking after his flock because he values sheep in themselves. He wants them for their wool and for their flesh. Likewise with crops: "And six years thou shalt sow thy land, and gather in the increase thereof; but the seventh year thou shalt let it rest and lie fallow, that the poor of thy people may eat; and what they leave the beast of the field shall eat. In like manner thou shalt deal with thy vineyard, and with thy olive yard" (Exodus 23:10–11). The intention here is to restore the fertility of the land so the owner can continue to exploit it. And this is so often the problem. Who is going to benefit – humans or animals and plants?

The Bible says: "Be fruitful and multiply and fill the earth and subdue it and have dominion over the fish of the sea and over the birds of the heavens and over every living thing that moves on the earth" (Genesis 1:28). Does one interpret "dominion" as lording over someone or something, or is it more a matter of caring for things in their own right? Canada is one of the Dominions of the British Commonwealth, as are Australia and New Zealand. I suspect they are inclined to think of themselves as equal to Great Britain – given the huge outflow, after the Second World War, from the mother country to the Dominions, perhaps more. In 1962, having finished my undergraduate degree in Britain, I scarpered to Canada. I certainly did not think I was downgrading myself. But if everyone is equal, that is hardly what the Bible says of the relationship of humans to animals. The English may think themselves top dogs. The English do think of themselves as top dogs. This is not a universal sentiment in the Commonwealth.

A more profitable line of criticism is that White is simply wrong to claim that it was the Judeo-Christian religion at the heart of the ecological crisis and more. In 1970, Lewis W. Moncrief denied that we have been fouling the nest since the start of time. All the evidence is that there was no such pollution. In any case, if there was pollution, Christianity is far from the sole culprit or the major culprit. White himself points out that humans were responsible for periodic flooding of the Nile River basin, and the use of fire as a method of hunting by prehistoric man has surely brought on significant "unnatural" changes in man's environment. Judeo-Christianity had no part in any of this (*Cultural Basis*, 509).

Of course, things have gone wrong, leaving us in the mess we are in now. That is another matter, and no cause for exaggeration. Moncrief lists three factors that have got us where we are today. First, obviously, urbanization and increased demand, raising the standard of living. "There is a high demand for goods and a need of services." Unfortunately: "The usual by-product of this affluence is waste from both the production and consumption processes. The disposal of that waste is further complicated by the high concentration of heavy waste producers in urban areas." Moreover: "the volume of such wastes is greater than the system can absorb and purify through natural means. With increasing population, increasing production, increasing urban concentrations, and increasing real median incomes for well over a hundred years, it is not surprising that our environment has taken a terrible beating in absorbing our filth and refuse" (510).

Second, a peculiarly American cause. The expansion West.

> To many frontiersmen, particularly small farmers, many of the natural resources that are now highly valued were originally perceived more as obstacles than as assets. Forests needed to be cleared to permit farming. Marshes needed to be drained. Rivers needed to be controlled. Wildlife often represented a competitive threat in addition to being a source of food. Sod was considered a nuisance – to be burned, plowed, or otherwise destroyed to permit "desirable" use of the land. (510)

And, third, technology. "It is very evident that the idea that technology can overcome almost any problem is widespread in Western society. This optimism exists in the face of strong evidence that much of man's technology, when misused, has produced harmful results, particularly in the long run." False optimism: "After all, we have gone to the moon. All we need to do is allocate enough money and brainpower and we can solve any problem" (511).

Who is right? White or his critics? One suspects that, as is often the case in disputes like this, there are elements of truth on both sides. It would be strange indeed if Christianity had no role to play in our present predicaments. If Christianity can tear America apart on the abortion issue, it must surely have played a role in such a major phenomenon as Global Warming. This said, it would be even stranger if things like urbanization and faith in technology had no role to play. Assuming that this is true, let's pull back and consider some of the more philosophical and religious implications.

Christianity – Are We Superior?

Plunge into an issue encountered earlier in this book. Wherein lies the source of value in the world, God or just nature itself? Start with what White was saying and why what he was saying was hardly that idiosyncratic. God exists and is good. He created humans and made us special, like Himself in fact. Animals and plants are there for our (human) good. Many others think the same way, for instance the secular thinker Edward O. Wilson, who – perhaps showing the influence of his evangelical childhood – wrote: "For if the whole process of our life is directed toward preserving our species and personal genes, preparing for future generations is an expression of the highest morality of which human beings are capable." Then, having established our status and needs, we learn: "It follows that the destruction of the natural world in which the brain was assembled over millions of years is a risky step" (*Biophilia*, 121). We ought to look after the Brazilian rainforests for they may harbor plants that, time will tell, have medicinal powers of vital use for the health of human beings.

It is not always easy to tease out people's mechanist versus organicist commitments. We saw reason in the last chapter to suggest that Wilson had significant organicist yearnings – his belief in progress, his endorsing group selection. Here, however, he seems to be thinking more in a mechanist-type manner, especially in the unquestioned assumption that the key to future wellbeing is going to depend on biotechnology, as in using plants from the rainforests to acquire substances that will be of value to us humans. This contrasts with a position that gives animals and plants value in their own right. It is true that, on the one hand, you might say that such a position is mechanist, because unlike organicism it does not privilege human beings. On the other hand, perhaps more importantly, it is organicist in that it takes a more integrated view of the living world, with all members having full value status. They are worthy parts of the whole. It is this kind of thinking that lies behind the sort of position promoted by New York philosopher Paul W. Taylor. He tells us: "From the perspective of a life-centered theory, we have prima facie moral obligations that are owed to wild plants and animals themselves as members of the Earth's biotic community." So, what follows from this? "We are morally bound (other things being equal) to protect or promote their good for *their* sake." All very Kantian. In the *Metaphysics of Morals*, one of the versions of the Categorical Imperative – the supreme rule of moral behavior – Immanuel Kant says that we should treat people as ends not as means. "Making an example of someone," to scare us into moral behavior, is wrong whatever the outcome.

> I maintain that man—and in general every rational being—exists as an end in himself and not merely as a means to be used by this or that will at its discretion. Whenever he acts in ways directed towards himself or towards other rational beings, a person serves as a means to whatever end his action aims at; but he must always be regarded as also an end. (28)

Taylor would extend this out from humans to animals and plants: "Our duties to respect the integrity of natural ecosystems, to preserve endangered species, and to avoid environmental pollution stem from the fact that these are ways in which we can help make it possible for wild species populations to achieve and maintain a healthy existence in a natural state" (*Ethics*, 197). But why do we have "moral obligations"? Why do we have such "duties"? "Such obligations are due those living things out of recognition of their inherent worth. They are entirely additional to and independent of the obligations we owe to our fellow humans." Taylor recognizes that often helping ourselves is also helping plants and animals. But the intentions are different. "Although many of the actions that fulfill one set of obligations will also fulfill the other, two different grounds of obligation are involved. Their well-being, as well as human well-being, is something to be realized *as an end in itself*."

Who is correct? Are humans unique in their possession of rights, or do these extend out through the living world? Is it simply silly – heretical – to say that animals and plants have rights in themselves? Are Taylor, and people like

Holmes Rolston III who speaks of non-humans as having "intrinsic natural values," truly speaking for Christianity? Many Christians would deny this. We have already quoted the Lutherans on the topic. Likewise, the evangelical Christian faculty at Wheaton College, Billy Graham's alma mater: "Scripture provides a logical value system. It establishes that the whole creation in general, and every part of it in particular, has a value given to it by God. This does not mean that the creation is inherently good or that it has the right to exist on its own merits, independent of God. Its goodness is derived from its Creator and so is a kind of 'grace' goodness, freely given in love, not grudgingly merited by right" (*Redeeming Creation*, 53). One is not saying that sparrows have no value. It is just that sparrows have no value in themselves. Their vanishing because of the lack of horse droppings must be related to how God feels about all of this. And how he feels is, unambiguously, right there in the first chapter of Genesis.

29 Then God said, "I give you every seed-bearing plant on the face of the whole earth and every tree that has fruit with seed in it. They will be yours for food.
30 And to all the beasts of the earth and all the birds in the sky and all the creatures that move along the ground – everything that has the breath of life in it – I give every green plant for food." And it was so.

There is not a lot of ambiguity here. One could say that humans have value because of God and so it is reasonable to think that other organisms have value because of God. This is surely true. "God created great whales, and every living creature that moves" (Genesis 1: 21). It still remains that God made them for our benefit.

Christianity – Is Technology Inherently Bad?

We are made in the image of God. Ask, as we have asked before: What does this mean? We have reasoning powers like (not necessarily equal to) God. We have a moral sense like God. Whatever the causes of Global Warming, we can surely say that it is the result of people using their reason. As Moncrief argues truly, whatever the end result, it is a product of our faith in technology. And if technology is not something produced by our power of reason, it is hard to know what would be. Moreover, note that for all that Global Warming may be a threat, it simply isn't the case that it all came about because we ignored God's moral dicta. Whatever the ill effects of the automobile, it has obvious moral virtues. If my appendix bursts, I want an ambulance to take me to hospital, not a horse and cart. Likewise, if the Green Revolution is not working in the image of God, it is hard to know what is. Making genetic alterations of plants is science of the highest order. In this respect the Green Revolution is deeply Christian. "And as ye would that men should do to you, do ye also to them likewise" (Luke 6:31).

If any one of us had a child who was threatened with blindness because of vitamin deficiency, would we not think that a scientist who set about finding a way to avoid this was showing Christian virtues of the highest kind? What about unfortunate side-effects? "High-yielding Varieties are more prone to pests and diseases compared with traditional cultivars, thus, requiring high level of pesticides," according to the International Rice Research Institute (*Great Rice Robbery*, 207). The obvious response is that we should do something about this. As things go wrong, or unexpected and unfortunate side-effects emerge, we set about dealing with them. Suppose, for instance, one is taking a medicine causing a significant improvement in health, but that it brings on diarrhoea. The solution is not to stop taking the medicine. It is rather to do something about the diarrhoea. (I speak here with some experience and passion. Six years ago, I was diagnosed with Idiopathic Pulmonary Fibrosis. The usual time span until death was two years. Thanks to a new drug, here I am, having just taken the dogs for a walk, writing this book. There are side-effects, but one deals with these, with as much good humor as one can summon. In my case, a fair amount is needed.)

God gave us the Earth and its contents. He gave us the powers to deal with them. After all, if, despite all our efforts and good intentions, we are not up to the challenge and if God doesn't like the way things are, He can change them, just as He can bring the whole show to an end, if He so decides. We can worry about that if and when it happens. For now, note that it is almost tautological to say that we are thinking within the machine root metaphor. If discovering how to use coal to drive steam-engines is not mechanical thinking, if messing around with the genes is not mechanical thinking, then it is hard to know what is. The same for attempts to put things right. As noted earlier, today in Florida – the "Sunshine State" – acre after acre is covered with solar paneling. It may be hard cheese on the miners of West Virginia, but pollution is being reduced drastically. Again, this all comes about from the use of reason – reason operating within the machine root metaphor. And before one starts to bemoan the fate of the miners, let's start by exploring the feasibility of opening factories producing electric-powered vehicles.

But isn't this the very problem? We are putting all the burden on God, so we are not to blame when things go wrong. This is a function of our thinking exclusively within the machine metaphor. No one will say there are no problems if we think holistically. Yet, if we were to start and stay within a picture of the whole, looking from the first at interactions good and bad, we have a much better chance of avoiding something like Global Warming. The poet Johann Wolfgang von Goethe knew what was what.

> Whatever Nature undertakes, she can only accomplish it in a sequence. She never makes a leap. For example she could not produce a horse if it were not preceded by all the other animals on which she ascends to the horse's structure as if on the rungs of a ladder. Thus every one thing exists for the sake of all things and all for the sake of one; for the one is of course the all as well. Nature, despite her seeming diversity, is always a unity, a whole; and thus, when she manifests herself in any part of that whole, the rest must serve as a basis for that particular manifestation, and the latter must have a relationship to the rest of the system. (*Goethe on Science*, 60)

It might seem that the Green Revolution, machine-thinking produced, is more successful. But as Vandana Shiva points out, this is not entirely clear cut. Organicism beckons. "Only in this way can we be enabled to preserve and respect the diversity of all life forms, including their cultural expressions, as true sources of our wellbeing and happiness" (*Ecofeminism*, 6).

Is any of this Christian, or at least an option for the Christian? Listen to Oberon Zell-Ravenheart (better known in his college days as Tim Zell): "it is a biological fact (not a theory, not an opinion) that ALL LIFE ON EARTH COMPRISES ONE SINGLE LIVING ORGANISM! Literally, we are *all* 'One'"(92); continuing: "The blue whale and the redwood tree are not the largest living organisms on Earth; the ENTIRE PLANETARY BIOSPHERE is." Do not be cowed by the fact that Zell-Ravenheart is an "initiate in the Egyptian Church of the Eternal Source" as well as "a Priest in the Fellowship of Isis." He is a Pagan! One might have a certain admiration for someone who has so firmly carved out his alternative lifestyle, especially since he has the good sense to live in Northern California rather than North Dakota. But is he the kind of person who should be a model for the Christian? I doubt polyamory is high on the papal want list. Nevertheless, Zell-Ravenheart and Pope Francis can sound remarkably similar in tone, thinking and speaking in an organic sort of way, where everything is integrated into a joint interdisciplinary approach. In 2015, the Pope issued an encyclical letter on caring for the environment, condemning what we humans have done and pleading for a change:

1. "Laudato si', mi' Signore" – "Praise be to you, my Lord." In the words of this beautiful canticle, Saint Francis of Assisi reminds us that our common home is like a sister with whom we share our life and a beautiful mother who opens her arms to embrace us. "Praise be to you, my Lord, through our Sister, Mother Earth, who sustains and governs us, and who produces various fruit with coloured flowers and herbs."

2. This sister now cries out to us because of the harm we have inflicted on her by our irresponsible use and abuse of the goods with which God has endowed her. We have come to see ourselves as her lords and masters, entitled to plunder her at will. The violence present in our hearts, wounded by sin, is also reflected in the symptoms of sickness evident in the soil, in the water, in the air and in all forms of life. (*Climate Change and Inequality*)

One presumes there is the (very important) difference that whereas for Zell-Ravenheart the value is inherent in the world, for the Pope this was value from God. Either way, what His Eminence writes could have come from the pen of Vandana Shiva, although to date the Pope seems not to have made the link she sees between organicism and feminist thinking.

Ecofeminism! The metaphor of Mother Earth is of very long standing. As this nineteenth-century poem by Emily Dickinson shows, the idea of the eternal feminine – giving and caring – is nigh commonsensical:

> She sweeps with many-colored Brooms—
> And leaves the Shreds behind—
> Oh Housewife in the Evening West—
> Come back, and dust the Pond!

(Poem 219)

With the 1960s rise of feminism generally, it was natural that attention should turn to environmental issues, with the predictable conclusion that it is men to blame for everything. The phallocentric nature of so much environmental thinking is a constant theme. According to Arisika Razak, "The physical rape of women by men in this culture is easily paralleled by our rapacious attitudes toward the Earth itself. She, too, is female. With no sense of consequence in the scant knowledge of harmony, we gluttonously consume and misdirect scarce planetary resources" (*Reweaving the World*, 165). As predictable is the linking of sexuality with organicism. Women's sexuality is an expression of the "essential, holistic nature of life on Earth," says Charlene Spretnak. Not to be overly modest, such expressions "are 'body parables' of the profound oneness and interconnectedness of all matter/energy, which physicists have discovered in recent decades at the subatomic level" (*Healing the Wounds*, 129).

A conservative thinker might think that this elevation of the female over the male is not necessarily more congenial to the Pagans than orthodox Christian thinking. Hold that thought for a minute. We will pick it up in the next chapter when we turn directly to prejudice against women. For now, let us leave matters. Like the rest of us, Christians have strong responses to both Global Warming and the Green Revolution. Like the rest of us, these responses are not uniform. What philosophy and theology tell us is that there is no single unambiguous Christian way of resolving disputes. Not to mention that there is the reasonable option of putting religion to one side and getting on with things in a secular manner.

6 Hate

The Problem

If we have a moral code that meshes with Christianity, why then are we so often non-social and why do we so blatantly disobey the dictates of Jesus Christ? "Love your enemies, bless them that curse you, do good to them that hate you, and pray for them which despitefully use you, and persecute you" (Matthew 5:43–4). Think about it. The Great War (later called the First World War), 1914–1918, between (depending how you count) 20 and 40 million dead. The Second World War, 1939–1945, 60 to 80 million dead. The Russian Civil War, 1917–1922, 5 to 10 million dead. The Chinese Civil War, 1927–1949, 10 million dead. And so it goes. We are not yet at the pogroms, from the Turks killing Armenians, Stalin and the Kulaks, Hitler and the Jews, down to Rwanda and the killing of the Tutsis, not to mention half a million women raped as a preliminary to grotesque mutilation of genitals.

The story continues. Think of America and the slaves. Thomas Jefferson. "The God who gave us life, gave us liberty at the same time." This is the man who had six children with his slave Sally Hemings and freed his children – 7/8 White – only on his death. At the time of the Revolution, there were 700 thousand slaves in America. At the time of the Civil War, there were four million slaves in America. In case you were wondering, when Christopher Columbus crossed the ocean, there were five million indigenous people living in North America. By the end of the nineteenth century, there were a quarter of a million. The British were not much better. In the famine of the 1840s, a million Irish people starved to death. Another million emigrated. True, it was not the British who invented or imported the potato blight. Business must not be stopped. "In the year A.D. 1846, there were exported from Ireland, 3,266,193 quarters of wheat, barley and oats, besides flour, beans, peas, and rye; 186,483 cattle, 6,363 calves, 259,257 sheep, 180,827 swine" (*Land Monopoly*, 10). At least enough food to feed half of the Irish population.

Why does such hatred occur? Why do we have prejudice – against foreigners, against those of other classes, against people of other colors, against homosexuals, against people of other religions, against Jews, against women? What is the biology of it all? What is the Christian response? Start with war.

The Darwinian Explanation

Natural selection is about surviving and reproducing. Getting yourself killed is not an advisable strategy. So how does war come about? Dig back into our history. For the past five million years, until recently, we were hunter-gatherers. All important is the fact that, comparatively, there were very few of us. A band or tribe could go about their own business. Of course, there would be some social, almost certainly sexual, intercourse between the groups, but if things started to turn nasty – too many men of the other group came after your women – you could simply get up and move away. No need for fighting – a conclusion confirmed by the archeological record. There was probably violence equally or exceeding that of American inner cities, but not systematic. More killing off the burdens – the handicapped, the elderly, and the like. No doubt fighting over women and so forth. But no war. R. B. Ferguson explains:

> True, in some cases, war could be present but not leave any traces. However, comparison of many, many cases, from all different regions, shows some clear patterns. In the earliest remains, other than occasional cannibalism, there is no evidence of war, and barely evidence of interpersonal violence. In Europe's Mesolithic [15,000–5,000 BCE], war is scattered and episodic, and in the comparable Epipaleolithic [20,000–8,000 BCE] in the Near East it is absent. (*Prehistory*, 191)

What happened? Simply, about 10 thousand years ago, agriculture came on the scene. At once there was a population explosion. For a farmer, children are much to be desired – working in the fields, minding the animals, and so forth. At the same time there were fixed assets. You cannot get up and take your farm with you. And with the population growth, people started to edge up to each other, and some had and some had not. A perfect recipe for war. If one turns to Hitler's 1925 manifesto, *Mein Kampf*, it is all there. "The foreign policy of a racial state has a duty to protect the existence of the race which forms the state on this planet by creating a natural, strong, and healthy relationship between the number and growth of the people and the quality of the soil and the size of the territory occupied. A healthy relationship only exists when the nutritional needs of a nation are met through its own territory and soil."

We are not innate killers. Attempts to prove otherwise are ineffective. Often cited are the findings of Napoleon Chagnon about violence between Amazonian tribes, the closest thing we have today to the hunter-gatherers of old. However, as critics have shown, Chagnon's subjects are anything but primitives. To the contrary; Ferguson again: "Ethnohistory and archaeology reveal large riverine settlements, bordering on urban scale, linked together with connections of trade, marriage, war, alliance, and ritual. These systems reached into smaller scale societies in highland interiors. Without any doubt, these were social worlds full of tumult, change, and conflict" (*History*, 382). The same failure of analogy occurs when appeals are made to the great apes. It is true that chimpanzees do go out on raiding parties, killing members of other groups. But, apart from the fact that humans have been evolving alone for five million years, compared to the great apes our natural weapons of killing are much reduced if not absent. No fearsome teeth, or other means to kill one another. In any case, the bonobos – pigmy chimps – are notorious for being straight out of the sixties. "Make love, not war."

Just War Theory

For many Christians – Mennonites, Quakers, Jehovah's Witnesses – the Sermon on the Mount is absolute. "Love your enemies, bless them that curse you, do good to them that hate you." What of the others, the majority, who feel that they must find some compromise between their faith and the need, the obligation, to support their country in war? The obvious move is to point out that the Bible is by no means unambiguous when it comes to war. Notwithstanding his sexual shenanigans, David was God's favorite, and this as much for his valor as for his verse making. In the New Testament, showing that things are not absolute, there is the story of the centurion who came asking Jesus to heal his servant. The centurion tells Jesus that his power is enough to do the task at a distance. "Lord, I do not deserve you to come under my roof" (Matthew 8:8). Jesus is touched. "Truly I tell you, I have not found anyone in Israel with such great faith" (8:10).

That said, the Bible is not greatly helpful in working out the details of acceptable war. No surprise that the leading Christian thinkers on the topic, Augustine and Aquinas, turned elsewhere, to the writings of the pre-Christian Roman, Cicero. Much is made of the distinction between *jus ad bellum* and *jus in bello*. Going to war without provocation falls under *jus ad bellum*. Offensive war, such as Hitler's march into Poland on September 1, 1939, is always condemned. It is defensive war, such as the Russian opposition to Napoleon's march on Moscow, that is allowable – in Augustine's opinion, morally obligatory. But then, conduct in the war, *jus in bello*, is equally important. Cicero: "Not only must we show consideration for those whom we have conquered by force of arms but we must also ensure protection to those *who* lay down their arms and throw themselves upon the mercy of our generals, even though the battering-ram has hammered at their walls" (*De Officiis*, in *War and Christian Ethics*, 29). The Russians raping the women in East Prussia as they entered Germany is a case of such immorality. So also, perhaps more controversially, is obliteration bombing, as the Allies inflicted on Germany as the Second World War progressed. The destruction of Dresden was wrong. In the opinion of many, not necessarily pacifist – philosopher Elizabeth Anscombe for a start – likewise the nuclear bombing of Hiroshima and Nagasaki.

One important thing is that, if the paleoanthropologists are right about the causal influence of agriculture, we are not innately – genetically – hardwired for war. If we were, short of genetic engineering, war could hardly be a moral issue. The man-eating tiger is a great threat and danger. In itself, this has nothing to do with morality. The tiger does what it does because it is a tiger. One is hardly going to condemn outright the war-like natures of many people who have been brought up in military-favoring environments; the Junkers in late nineteenth-century East Prussia, for instance. But there is a dimension of freedom. The My Lai incident in Vietnam, when American soldiers raped and murdered 500 villagers, was morally wrong. The leader of the massacre, Lieutenant William Calley, should have known better. The US military was sensitive to, and took account of, such appalling incidents, and much subsequent training has been designed to avoid their repetition. No one takes the issues lightly.

The Organismic Perspective on War

If one (properly) thinks of Darwinism under the root metaphor of the machine, what of approaches to war from under the root mechanism of the organism? General Friedrich von Bernhardi, pushed out of the German army because he was signaling a little too bluntly the General Staff's intentions, left no place for the imagination in his best-selling *Germany and the Next War*, published in 1912. "War is a biological necessity," and hence: "Those forms survive which are able to procure themselves the most favourable conditions of life, and to assert themselves in the universal economy of nature. The weaker succumb." Progress depends on war: "Without war, inferior or decaying races would easily choke the growth of healthy budding elements, and a universal decadence would follow." And, anticipating horrible philosophies of the twentieth century: "Might gives the right to occupy or to conquer. Might is at once the supreme right, and the dispute as to what is right is decided by the arbitrament of war. War gives a biologically just decision, since its decision rests on the very nature of things" (quoted by Crook in *Darwinism, War and History*, 83).

Von Bernhardi knew the real opponent of Germany – Britain! Magnanimously in a book published in 1914, at the beginning of the war – *Our Future: A Word of Warning to the German Nation*, which the translator somewhat inventively rendered as *Britain as Germany's Vassal* – he set out the conditions needed to avoid war between the two countries. For a start: "England would have to give Germany an absolutely free hand in all questions touching European politics, and agree beforehand to any increase of Germany's power on the Continent of Europe which may ensue from the formation of a Central European Union of Powers, or from a German war with France" (152). From Elizabethan times or earlier, British diplomacy had centered on not letting any single European state dominate all the others. Von Bernhardi obviously knew that his suggestions would not be taken seriously.

Germany must arm itself to the hilt. War is coming, no matter what we do. Take note of the appeal to the organic nature of the state and how it leads to struggle with others. "Every nation possesses an individuality of its own, and all progress among nations is based on their competition among themselves. As the competition among nations leads occasionally and unavoidably to differences among them, all real progress is founded upon the struggle for existence and the struggle for power prevailing among them." This is all for the good. "That struggle eliminates the weak and used-up nations, and allows strong nations possessed of a sturdy civilisation to maintain themselves and to obtain a position of predominant power until they too have fulfilled their civilising task and have to go down before young and rising nations" (26). Regretfully, came the war and von Bernhardi's dog was licked. The old general was unregenerate. Germany's defeat came entirely through mismanagement. Von Bernhardi's philosophy of struggle and conflict was unchanged. "Always, as long as humans are humans, force in its most encompassing sense will indicate the political and cultural meaning of states. It is at the root of all mental and moral progress" (*Vom Kriege*, 237). Note the yearning for progress. War is not a good in its own right. It is rather that it leads to the desired kind of change – upwards.

Not a great deal of von Bernhardi's overview comes directly – or indirectly – from evolutionary thinking, especially not Darwinian thinking. One would hardly expect this. He despised the British. Friedrich von Bernhardi's ideas owe little to British industrialism and much to German idealism. For Hegel and successors, not only was war a necessary and inevitable thing, in respects it was a good thing for us morally. War is a necessary component in defining or delimiting one state from another. Above all, this was shown by the growth and coming together of the parts to make the new Prussian-dominated Germany. "I have remarked elsewhere, 'the ethical health of peoples is preserved in their indifference to the stabilisation of finite institutions; just as the blowing of the winds preserves the sea from the foulness which would be the result of a prolonged calm, so also corruption in nations would be the product of prolonged, let alone "perpetual" peace'" (*Elements*, 324; Hegel's jab was directed against Kant, who had spoken of the possibility of perpetual peace).

Others were in tune. Thus, the philosopher Fichte: "The noble-minded man's belief in the eternal continuance of his influence even on this earth is thus founded on the hope of the eternal continuance of the people from which he has developed ... Hence, the noble-minded man will be active and effective, and will sacrifice himself for his people" (*Addresses*, 135). As opposed to the kind of thinking we find in Darwin, although to be found in the Germanic-influenced Spencer, the value of the group over the individual is prominent. "Life merely as such, the mere continuance of changing existence, has in any case never had any value for him; he has wished for it only as the source of what is permanent. But this permanence is promised to him only by the continuous and independent existence of his nation. In order to save his nation, he must be ready even to die that it may live, and that he may live in it the only life for which he has ever wished" (136).

The holistic thinking of Hegel and others was magnified during the nineteenth century, especially by the Volkish movement that nigh-idolized a glorious (albeit fanciful) medieval past, well captured and romanticized by the fairy tales collected by the Grimm brothers and, dramatically, by the operas of Wagner, with their epic stories of humans and gods in a now-vanished world.

> Honour your German Masters,
> then you will conjure up good spirits!
> And if you favour their endeavours,
> even if the Holy Roman Empire
> should dissolve in mist,
> for us there would yet remain
> holy German Art!
>
> (Wagner, *Die Meistersinger von Nürnberg*)

We have a vision of the state as an organic unity, represented today by the German state. A state on its way up. "In the extrasocial struggle, in war, that nation will conquer which can throw into the scale the greatest physical, mental, moral, material, and political power, and is therefore the best able to defend itself," to cite von Bernhardi (*Germany and the Next War*). There is progress, but not British industrial progress forged by sweat and intelligence – rather, the progress of the World Spirit, the progress of German idealism.

Christian Response

Whether or not one accepts it, no one could deny that the Darwinian explanation of war is firmly scientific. It could be wrong, but the Popperian would point out that that is the whole point. One might perhaps feel less confidence in saying that this German idealistic account of war is truly scientific, as opposed to metaphysical or some such thing. It is not obvious what would or could falsify it. One certainly can say that it does not mesh well with the

Darwinian account, finding a propensity for war in human nature, as opposed to the Darwinian denial. In that sense, at least, right or wrong, it is scientific. Hegel demanded strong restrictions about not harming civilians, so there is certainly some awareness of Just War Theory, but generally the philosophy is different. We are facing "Holy War." Just War thinking starts with the premise that there are disinterested moral rules to be obeyed. In Holy War thinking, anything goes so long as it is in the supposed interests of God and the religion in which He is embedded. "Not just independently but repeatedly and centrally, official statements and propaganda declare the war is being fought for God's cause, or for his glory." To this end, says Philip Jenkins, there is identification of "the state and its armed forces as agents or implements of God" (*Great and Holy War*, 6). It was in the Great War, the First World War, that this philosophy or theology – Holy War – got expressed most clearly. "German Christianity represents the right relation between Christ and His disciples, and our nature the most perfect consummation of Christianity as a whole. We fight, then, not only for our land and our people; no, for humanity in its most mature form of development; in a word, for Christianity as against degeneration and barbarism" (*Hurrah and Hallelujah*, 69). And so on and so forth, at length.

If the Christian has trouble in the face of Just War Theory, what can possibly be said in the face of Holy War? One possible move might be to stress the significance of original sin. Thanks to Adam and Eve and the forbidden apple eating, we are all born with a propensity for sinning. I am not sure that someone like von Bernhardi would accept that going to war for one's nation, particularly if that nation happens to be Germany, would count as sinful; but it does perhaps open the door for the Christian to say that going to war, for whatever reason, is part of human nature. It is not something added on contingently, as would seem to be the violent after-effects of agriculture. Is there any empirical evidence to back up this hypothesis? Many have thought, still think, that there is. As was made terrifyingly clear in the prologue to the movie *2001*, we are killer apes. Appearances apart, for all the surface skin of sociality, we humans are evil to the core (Fig. 6.1).

Paleoanthropologist Raymond Dart was the modern originator of this idea. Rightly celebrated as the person who in 1924 identified Taung Baby, *Australopithecus africanus* – a human ancestor ("hominin") – by the 1950s Dart was trying his hand at causal explanations. At the risk of making an awful

Figure 6.1   WWI poster – killer ape.

pun, he made no bones about his views. We were carnivores. Worse, we were not that choosy about whom we ate. For whatever reason – taste or necessity – we ate each other. That explains a lot: "The loathsome cruelty of mankind to man forms one of his inescapable characteristics and differentiative features; and it is explicable only in terms of his carnivorous, and cannibalistic origin" (*Predatory Transition*, 203). Warming to his theme:

> The blood-bespattered, slaughter-gutted archives of human history from the earliest Egyptian and Sumerian records to the most recent atrocities of the Second World War accord with early universal cannibalism, with animal and human sacrificial practices of their substitutes in formalized

religions and with the world-wide scalping, head-hunting, body-mutilating and necrophiliac practices of mankind in proclaiming this common bloodlust differentiator, this predaceous habit, this mark of Cain that separates man dietetically from his anthropoidal relatives and allies him rather with the deadliest of Carnivora. (208–9)

Dart was read eagerly by Robert Ardrey, an American film-script writer who, having run afoul of the McCarthy anti-communist hearings, moved to Africa where, making full use of his talents, he became a best-selling writer about human origins, pushing a view of humankind – *African Genesis* – that he admitted openly (and proudly) he got from the often-derided Dart. Like those early hunter-gatherers discovering the virtues of being bipedal, this was a story with legs. In some quarters it is still going strong today – notably in the writings of academic superstar Steven Pinker. We learn that we are "wired for violence," even if there is never occasion to show this. True, in conflict situations we often show restraint, some of us more than others. Don't kid yourself that this means we are nice underneath (*Better Angels*, 483). Pure and simple. We are nice only out of prudence.

In a way, this all seems a matter of we have lemons, let's make lemonade. Conflict leads to progress. How can we guarantee that there will be conflict? We are killer apes! You might object that the links to Christianity, even for those who accept the Fall and consequent original sin, are tenuous and somewhat forced. Dart thought otherwise. He starts his piece on man-as-a-killer-ape with a theological reflection. "Of all beasts the man-beast is the worst, To others and himself the cruelest foe," quoting *Christian Ethics* by Richard Baxter. Compare Zechariah 8:10: "For before those days there was no wage for man or any wage for beast, neither was there any safety from the foe for him who went out or came in, for I set every man against his neighbor." Remember also: "this common bloodlust differentiator, this predaceous habit, this mark of Cain that separates man dietetically from his anthropoidal relatives."

One wonders by this stage if we are getting the wisdom of the Bible rather than the wisdom of the rocks. Without digging further into this seam, perhaps we should bring the discussion to an end with a little cold water. Accounts of actual warfare strongly suggest that the urge and ability to kill our fellow humans is very far from being confirmed. For most of us such an urge is non-existent. "During World War II U.S. Army Brigadier General S. L. A. Marshall asked average soldiers [whom he was interviewing] what it was they did in battle" (Grossman, in *On Killing*, 3). His findings were staggering: "of every hundred men along the line of fire during the period of an encounter, an average of only 15 to 20 'would take any part with their weapons.' This was consistently true 'whether the action was spread over a day, or two days or three.'" Marshall stressed that the non-firing soldiers were not wimps or cowards – the very opposite, in fact. On asking those who had recently been in close combat with either European or Japanese foes, "The results were consistently the same: only 15 to 20 percent of the American riflemen in combat during World War II would fire at the enemy." However: "Those who would not fire did not run or hide (in many cases they were willing to risk great danger to rescue colleagues, get ammunition, or run messages), but they simply would not fire their weapons at the enemy, even when they were faced with repeated waves of banzai charges" (3–4). So much for killer apes. Without now claiming that it is impossible to reconcile organismic thinking about war with Christianity, we are but part way on a long and somewhat rocky road.

Prejudice

War is hatred, conflict, violence between groups. Prejudice is hatred, conflict, violence between individuals. Begin with some science. Gordon Allport, in his classic study of the issue – *The Nature of Prejudice* – defined it as "an avertive or hostile attitude toward a person who belongs to a group, simply because he belongs to that group, and is therefore presumed to have the objectionable qualities ascribed to that group" (8). Allport's emphasis is on hostility and dislike of the "outgroup." We are "ingroup"; they are outgroup. As it happens, today – especially in the light of our long history as hunter-gatherers – there tends to be more emphasis on ingroup. At least, it is ingroup where we start and which is our basis. Then we extend to outgroup. Thinking this way, what then is the importance of ingroup? Simply, we are living in small bands of people, about 50, whose chief characteristic, whose over-whelmingly important biological adaptation, is that we are social. Ingroup over outgroup. Cooperation rather than strength. Social learning rather than instinct. Exactly Darwin's sentiments: "No tribe could hold together if murder, robbery, treachery, &c., were common; consequently such crimes within the limits of the same tribe 'are branded with everlasting infamy'" (*Descent of Man* 1, 93). The very evolutionary position we have been promoting. "Ingroup membership is a form of contingent altruism. By limiting aid to mutually acknowledged ingroup members, total costs and risks of nonreciprocation can be contained.

Take note how nothing yet speaks to outgroup discrimination. In the words of Brewer, we are dealing with "relative favoritism toward the ingroup and the absence of equivalent favoritism toward outgroups. Within this framework, outgroups can be viewed with indifference, sympathy, even admiration, as long as intergroup distinctiveness is maintained" (*Psychology of Prejudice*, 434). This said, it is easy to see how outgroup aggression might be generated. Discrimination might come about as a by-product of ingroup favoritism. "Ultimately, many forms of discrimination and bias may develop not because outgroups are hated, but because positive emotions such as admiration, sympathy, and trust are reserved for the ingroup and withheld from outgroups" (439). As we move towards modern societies, we can see that this process might be of increasing importance.

Turn now to actual instances of prejudice. I shall select items of immediate importance and controversy today.

Despised Outgroups

Foreigners and Immigrants

Few recent acts equal the callous cruelty of Governor Ron DeSantis of Florida, who, in September 2022, in pursuit of the Republican nomination for president in 2024, shipped an unknowing planeload of Venezuelan immigrants from Florida up to Martha's Vineyard, in New England, on the false promise of homes and jobs. Many others compete for his title. Not to show undue prejudice, let me turn to the country of my birth and childhood, Great Britain. To borrow a phrase from Jane Austen, it is a truth universally acknowledged that the British have never much liked foreigners. Charles Dickens, in his novel *Little Dorrit*, captured the attitude. Writing of people in a working-class part of London, he noted how it "was uphill work for a foreigner" to make genuine contact with them.

In the first place, they were vaguely persuaded that every foreigner had a knife about him; in the second, they held it to be a sound constitutional national axiom that he ought to go home to his own country. They never thought of inquiring how many of their own countrymen would be returned upon their hands from divers parts of the world, if the principle were generally recognised; they considered it particularly and peculiarly British. In the third place, they had a notion that it was a sort of Divine visitation upon a foreigner that he was not an Englishman, and that all kinds of calamities happened to his country because it did things that England did not, and did not do things that England did. (302)

An attitude that persists to this day. Brexit did not come by chance. Being in the European Union meant that foreigners could come, without needing permission, to live and work in Britain. This was not liked. Listen to Nigel Farage, the leader of the Brexit movement, on the night (in June 2016) when the British voted to leave the Common Market (having joined in January 1973):

I've seen the people gathering in party mood and yes flying the Union Jack. And you know why? Because that's our national flag. We don't want the European star-spangled banner. We don't want their anthem. We don't want their president. We don't want their army. We, in six hours' time, are going to be free of this. And it is because of that that this is the greatest day in modern British history.

America keeps its end up. Donald Trump was elected President in 2016 in large part because he was anti-immigrant. He said, as reported by the *Washington Post*: "When Mexico sends its people, they're not sending their best. They're sending people that have lots of problems, and they're bringing those problems with us. They're bringing drugs. They're bringing crime. They're rapists." Comments political journalist Thomas B. Edsall: "A lot of Americans are susceptible to the kinds of rhetoric that won Trump the presidency: especially his appeals to people's innate xenophobia and fears of threats both internal and external." The most obvious symbol of Trump's stance against immigrants was the demand that a huge wall be built between America and Mexico; but there are many other things that he enacted to cut the number of immigrants and to return many of them to their own countries. As an aside, my state of Florida earlier this year enacted a law that made immigration that much more difficult. Someone who takes an unregistered child to sanctuary, perhaps a church, is breaking the law. One presumes now, having now blocked that influx of people who do the backbreaking work that Americas eschew, that the politicians will be helping with the picking of fruit and vegetables.

Paralleling Britain, America has a tradition. Immigration has always been a contentious issue. These vile attitudes did not come from nowhere. As David Reich explains, the new technique of ferreting out ancient DNA reveals that for a very long time our ancestors had been leaving Africa for Europe and Asia – up to two million years ago. Move down to the most recent significant move out of Africa, some 50 thousand years ago or a bit earlier. Those who went west to Europe got caught in an Ice Age, making much of Northern Europe uninhabitable. People were squashed down to places like Spain, beyond the grip of the glacial ice. While you might not want to go to war with your competitors, there would be increasing pressure to keep your distance and not let others grab or move in on what you now had – and conversely. An edge against outsiders would be of selective value – something that would not necessarily vanish as the ice receded and people could start to move north. The paradox is that we all became immigrants in one way or another. Around five or so thousand years ago, there was a major invasion from the east into Europe, displacing the then-inhabitants. Eventually this made its way across to the British Isles. The "Bell Beaker culture" – appropriately named after the invaders' style of pottery – arrived rather more than four thousand years ago. Genetic evidence – studies of ancient DNA – suggests that the newcomers really pushed aside the already-established inhabitants of the isles (see Fig. 6.2), with ongoing consequences for our heredity: "The genetic impact of the spread of peoples from the continent into the British Isles in this period was permanent" (*Who We Are*, 115). One senses hypocrisy in modern attitudes.

Race

America has the most appalling record of prejudice against Black people, African Americans. Years of slavery (Fig. 6.3.) Remember: from the time of the Revolution, towards the end of the eighteenth century, until the beginning of the Civil War in 1860, America's slave population grew from 700 thousand to over four million. And then, after that, a century of Jim Crow, as

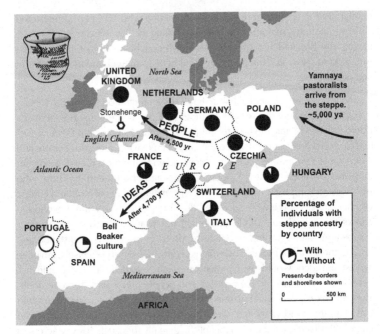

Figure 6.2   Arrival in Britain of the Bell Beaker culture.

Black people were regarded and treated as inferior. Attitudes persist. In 2022, the State of Florida, one of the most populous in the United States, has enacted one law after another making it difficult for Black people to vote. It has, for instance, barred Sunday voting, simply because that is a major time for Black voting: after morning service, guided by the pastor, into the church bus and on the way to the polling station. Not that this makes a huge difference. Thanks to redistricting, forced on the legislature by the governor (the same Ron DeSantis), in the forthcoming federal election, the number of Black congressmen will drop from four to two. It was not until earlier this year that the high school, in the district where my daughter lives, changed its name from that of the treasonous leader of the Confederate army, Robert E. Lee. Seventy percent of the students are Black.

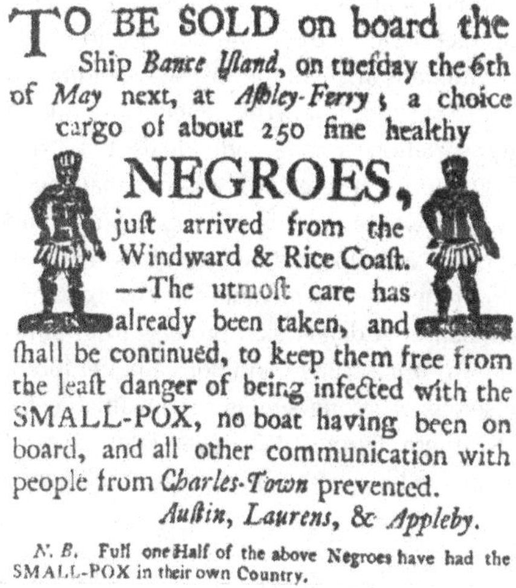

Figure 6.3 Advertisement for slaves.

Americans are not alone in being racist. "When civilised nations come into contact with barbarians the struggle is short, except where a deadly climate gives its aid to the native race. Of the causes which lead to the victory of civilised nations, some are plain and some very obscure. We can see that the cultivation of the land will be fatal in many ways to savages, for they cannot, or will not, change their habits" (*Descent of Man* 1, 239). And so on and so forth. Yet as we shall see repeatedly, Darwin may have been caught in the prejudices of the Victorian era. His theory had the seeds that led to their refutation. Most pertinently – natural selection! Where is the evidence that some races/groups are more or less intelligent than others? Even before you get into psychological testing and the like, the biology is against it. To speak of a race or subspecies, you need much more genetic differentiation than one finds between local populations. According to Smith et al., generally there is a 25% genetic variability demanded to speak of different "races" (*Subspecies and Classification*). Incredibly striking are the differences between chimpanzees and humans. In the case of the apes, the empirical evidence "confirms the reality of race in chimpanzees using the threshold definition, as 30.1% of the genetic variation is found in the among-race component ... In contrast to chimpanzees, the five major 'races' of humans account for only 4.3% of human genetic variation – well below the 25% threshold. The genetic variation in our species is overwhelmingly variation among individuals (93.2%)." *Homo sapiens* went through bottlenecks. There simply is not that much genetic variation in our species. And where there is, reasons are not hard to find. The best-known – notorious – variation is difference in skin colors. It is a function of the distribution of the pigment melanin, and its existence makes perfectly good Darwinian sense. A darker skin protects from ultraviolet radiation, a bigger problem in Africa. White skin does a better job of vitamin D synthesis, in the absence of strong sunlight – invaluable for those ongoing dark days of Northern Europe. Whatever later cultural overlays there may be, we are not talking here about brute intelligence or anything like that.

Sexual Orientation

The State of Florida has just passed laws severely circumscribing the teaching about such orientation in publicly funded classrooms. The Disney Corporation has a massive presence in Florida. The state has taken them on because they have, publicly, expressed support for people of minority sexual orientation and identity. Darwin, one suspects, would be happy with this. His science forced him to acknowledge homosexual behavior. He accepted the theory, promoted by the Scottish anatomist Robert Knox, of "primordial hermaphroditism." The primeval ancestor of all animals was both male and female. To this day, as Darwin wrote in a private notebook, all human beings show the traces of both sexes – "every man and woman is hermaphrodite" (D 162). This transcribed into views about behavior, and, in *The Descent*, Darwin acknowledged that homosexual orientation is something we are stuck with. But restricted to savages! "The greatest intemperance with savages is no reproach. Their utter licentiousness, not to mention unnatural crimes, is something astounding" (*Descent of Man* 1, 96). As with race, this proved the thin end of the wedge. By century's end, people were speculating about the naturalness of being homosexual. Darwin enthusiast and English naturalist Edmund Selous noted same-sex activity in birds, arguing that this "might even help to guide us in the wide domain of human ethics" (182). "Human ethics"! If homosexuality is natural, then should we condemn it as immoral? Little wonder that Sigmund Freud counted *The Descent* as one of his top books.

Women

A source of prejudice if ever there was one. Once again, *The Descent* does the honors. "Man is more courageous, pugnacious, and energetic than woman, and has a more inventive genius." That tells it like it is! "Male and female children resemble each other closely, like the young of so many other animals in which the adult sexes differ; they likewise resemble the mature female much more closely, than the mature male. The female, however, ultimately assumes certain distinctive characters, and in the formation of her skull, is said to be intermediate between the child and the man" (2, 317). Once again, science has taken Darwin's theory and shown how very wrong he is. In hunter-gatherer clans, the sexes were equal. They had to be. To use an unfortunate but appropriate metaphor, there was no place for weak sisters. The males might be working out the predicted path of their prey. The females were designing and making sophisticated traps to capture smaller animals. It was the coming of agriculture that changed all of that. Women became baby machines, a culture-fueled happening. And this being so, as we have seen over the last century or so, women have started to regain equal place. Labor-saving devices such as washing machines have released them from much day-to-day drudgery. Efficient contraception means that the baby-machine days are over. With predictable consequences. As psychologists Stephen Ceci and Wendy Williams argue:

> Roughly half the [American] population is female, and by most measures they are faring well academically. Consider that by age 25, over one-third of women have completed college (versus 29% of males); women outperform men in nearly all high school and college courses, including mathematics; women now comprise 48% of all college math majors; and women enter graduate and professional schools in numbers equal to most, but not all fields (currently women comprise 50% of MDs, 75% of veterinary medicine doctorates, 48% of life science PhDs, and 68% of psychology PhDs). (5)

This has not stopped the good State of Florida having in this year (2022) enacted laws further restricting abortion. The permissible period is now 15 weeks (available only after a 24-hour period of reflection). No exceptions for rape or fetal deformity or threats to the mother's life.

Christianity

Prejudice exists, but biology shows that, although perhaps understandable, it is not inevitable, and we can set about improving things. Where stands Christianity in all of this? Here we encounter one of the more interesting and important points in this whole book. Historically, Christianity has been far from a defender of the despised and weak and needy. Not just historically, if you look at the American South. Take our categories one by one.

Foreigners and Immigrants

> When the Lord your God brings you into the land you are entering to possess and drives out before you many nations—the Hittites, Girgashites, Amorites, Canaanites, Perizzites, Hivites and Jebusites, seven nations larger and stronger than you—and when the Lord your God has delivered them over to you and you have defeated them, then you must destroy them totally. Make no treaty with them, and show them no mercy.   (Deuteronomy 7:1)

This means no intermarriage, or anything else that might be construed as friendly. "Break down their altars, smash their sacred stones, cut down their Asherah poles and burn their idols in the fire." You are special. God's "treasured possession." There is no more to be said.

It is true that there are sentiments pointing the other way, starting with the Parable of the Good Samaritan. In the Old Testament too:

> 33 When an alien resides with you in your land, you shall not oppress the alien. 34 The alien who resides with you shall be to you as the citizen among you; you shall love the alien as yourself, for you were aliens in the land of Egypt: I am the Lord your God. (Leviticus 19)

One might add that, earlier in this chapter, we are given warnings that look very much as if sex with a slave is a lesser moral evil than sex with a free woman.

> [20] If a man has sexual relations with a woman who is a slave, designated for another man but not ransomed or given her freedom, an inquiry shall be held. They shall not be put to death, since she has not been freed; [21] but he shall bring a guilt offering for himself to the Lord, at the entrance of the tent of meeting, a ram as guilt offering.

Apparently, this lets him off the hook.

The conclusion is that the Bible is not consistent on the subject of foreigners and immigrants. You are going to have to use outside help to make a decision, and one presumes this means science.

Race

Race is intimately bound up with slavery, and we have just had intimation that the Bible does not regard slaves as on the same social footing as free persons. This is a consistent theme. Traditionally, the sin of Ham in looking at and laughing at his father's drunken nakedness is the reason why Noah cursed Canaan, Ham's son. This supposedly led to his descendants being Black and slaves. I am not sure how much credence should be given to this story. Wisely, the Creation Museum in Northern Kentucky – climb aboard the Ark! – stays safely away from it. More is surely merited in the way that Abraham and Sarah used her slave Hagar as a substitute when Sarah proved infertile and then cast her (and her son Ishmael) out into the wilderness when Sarah did in fact get pregnant (with Isaac). I leave without comment the claim by Christopher Clark that Sarah's giving Hagar to Abraham was one of "the great unselfish acts of her life" (*Iron Kingdom*).

Well known is the fact that not only did Paul acknowledge slavery, he endorsed it.

> [5] Slaves, obey your earthly masters with respect and fear, and with sincerity of heart, just as you would obey Christ. [6] Obey them not only to win their favor when their eye is on you, but as slaves of Christ, doing the will of God from your heart. [7] Serve wholeheartedly, as if you were

serving the Lord, not people, [8] because you know that the Lord will reward each one for whatever good they do, whether they are slave or free. (Ephesians 6)

It is little wonder that the slave owners of America made much of the Christian backing for their position. My argument today is obviously not that they were right, but that it is our knowledge of science that corrects the misunderstanding, not the other way around.

Sexual Orientation

The same story over again. Old Testament: "You must not have sexual intercourse with a male as one has sexual intercourse with a woman; it is a detestable act" (Leviticus 18:22). New Testament: Paul tells us of dreadful people. "For although they knew God, they did not glorify him as God or give him thanks, but they became futile in their thoughts and their senseless hearts were darkened." God was not pleased.

> [26] For this reason God gave them over to dishonorable passions. For their women exchanged the natural sexual relations for unnatural ones, [27] and likewise the men also abandoned natural relations with women and were inflamed in their passions for one another. Men committed shameless acts with men and received in themselves the due penalty for their error. [28] And just as they did not see fit to acknowledge God, God gave them over to a depraved mind, to do what should not be done. (Romans 1)

Again, it is open to the Christian to disregard the literal text and to go with science. We know now that homosexuality is not a perversion, certainly not a vile way of life freely chosen and dangerous to others. Love is what matters, not sexual desires and activities. No need to mention that Jesus and his disciples are hardly a guide to heterosexuality. Or perhaps it should be mentioned. I have long publicly asserted my conviction that Jesus was gay. Such a hypothesis makes sense of so much, from Jesus' somewhat colorless relationship with Joseph his father and his rather intense relationship with Mary his mother, to the unambiguous references to John as his favorite disciple: "There was reclining on Jesus' breast one of His disciples, whom Jesus loved" (John 13:23). Above all the total lack of any sexual interest in women. Jesus was friends with Mary and Martha, but there is absolutely no suggestion of any sexual attraction or behavior. Jesus cared about other people, and that was enough for him as it is enough for me. "Love is what matters."

Women

St Paul again. There is a proper ordering and women are not first.

> —as in all the congregations of the Lord's people. Women should remain silent in the churches. [34] They are not allowed to speak, but must be in submission, as the law says. [35] If they want to inquire about something, they should ask their own husbands at home; for it is disgraceful for a woman to speak in the church. (1 Corinthians 14)

No comment, or rather lots of comment. Picking up discussion started in the last chapter, it is open to the Christian to argue that, although Paul's comments may have been acceptable in his time, his time is not our time. Given what we have learnt about sexual differences, this kind of discrimination is without warrant and should be dropped. At least, that is the liberal position on all this and the other instances of prejudice. Do not be deceived that all people are going to take this position. As noted above, in the year 2022 the State of Florida has enacted laws against immigration, supporting racial prejudice, "protecting" children and others from the nature and views of minority sexual orientations, and imposing further restrictions on the freedom of women to have lives of their choice. The support for these positions and actions comes nigh exclusively from people who think these are mandated by their religion – usually conservative Catholic or evangelical Protestants. One does not need one hand to count the number of female Roman Catholic priests. The Southern Baptists are little different. According to the Southern Baptist Convention Executive Committee, "The vast majority of Southern Baptist congregations call only men to serve as pastor. About 30 of over 40,000 churches currently have a woman as pastor – only 2 of 5,000 Southern Baptist congregations in Texas." Thank you, St Paul: "I do not permit a woman to teach or to have authority over a man" (1 Timothy 2:12). Naturally, there are responses. Apollos, a native of Alexandria, came to Ephesus. "He began to speak boldly in the synagogue. When Priscilla and Aquila heard him, they invited him to their home and explained to him the way of God." Apparently, one does not have to wait until the twenty-first century to find learned women prepared to correct overly self-satisfied men.

No discussion today would be complete without a discussion on abortion, especially with respect to conservative states in the South and mid-West. Now, with the overturn of Roe vs Wade, vast swaths of America are moving toward the position of Poland, where abortion is banned from the moment of conception. With predictable consequence, as journalists Katrin Bennhold and Monika Pronczuk explain:

> Ms. Sajbor, a 30-year-old hairdresser from the small town of Pszczyna in southern Poland, had been thrilled to be pregnant. She wanted a sibling for her 9-year-old daughter, Maja.
>
> It was a shock when in her 14th week she learned that doctors suspected her fetus had Edwards syndrome, a serious chromosomal abnormality, and diagnosed other malformations. Instead of a nose, there was only cartilage. The feet were deformed. One heart chamber was dysfunctional.

The baby would be born dead, or so seriously deformed it would die within a year. No abortion. Then her waters broke. She was taken to hospital. No abortion.

> By the time the baby's heartbeat had stopped and the doctors took Ms. Sajbor into the operating room the next morning, her limbs had already gone blue.
>
> She died at 7:30 a.m.

In Poland, it is the conservative Catholic Church that is pulling the strings. In America, the Church is joined by Protestant Evangelicals. Neither of them justify the belief that the opposition is theological, rather than purely political, keeping women in their place. The Catholic Church was always anti-abortion as a general practice, but far from absolute. Aristotle had said that male fetuses only quicken after 40 days, whereas females took 80 days. But then in the nineteenth century, the Immaculate Conception of Mary – she was born without original sin – was made dogma, and a general ban was soon forthcoming. Evangelicals, remarkably, even in the 1970s, accepted abortion, in some cases at least. From a Southern Baptist Convention statement of 1971: "Be it further resolved, that we call upon Southern Baptists to work for legislation that will allow the possibility of abortion under such conditions as rape, incest, clear evidence of severe fetal deformity, and carefully ascertained evidence of the likelihood of damage to the emotional, mental, and physical health of the mother." Then the Supreme Court ended the tax-free status of private schools founded simply to avoid segregation. Evangelicals looked for another cause they could promote, they put aside their anti-Catholicism which made them open to abortion, and the rest, as they say, is history.

This history hardly exonerates Christianity, but it does suggest that blanket, Richard-Dawkins-type condemnations are inappropriate. Christians take different positions on the status of women in general and on abortion in particular. It is left as an exercise for the reader if, when we have disputes like this, Christians might look towards science to guide them in their thinking.

Is Organicism Superior?

One doubts that the average organicist is going to disagree with much that has just been said. It is worth noting, though, that often there is a presumption that the mechanistic approach, specifically the Darwinian approach, is going to be somewhat soulless. Not really concerned about right or wrong, feelings or otherwise. The historian Caroline Merchant argues this in her deservedly praised book, *The Death of Nature*. She sees, in the move from organicism to mechanism, the perhaps-unintended consequence of downgrading the nature and role of women. We start with nurturing Mother Earth, and then mechanism kills her. "Between the sixteenth and seventeenth centuries the image of an organic cosmos with a living female earth at its center gave way to a mechanistic world view in which nature was reconstructed as dead and passive, to be dominated and controlled by humans" (xvi). This at once reflects on the status of women. "If women overtly identify with nature and both are devalued in modern Western culture, don't such efforts work against women's prospects for their own liberation?" (xvi). Expectedly, Darwinism has a role in this sad story. "In the nineteenth century, Darwinian theory was found to hold social implications for women. Variability, the basis for evolutionary progress, was correlated with a greater spread of physical and mental variation in males. Scientists compared male and female cranial sizes and brain parts in the effort to demonstrate the existence of sexual differences that would explain female intellectual inferiority and emotional temperament" (162–3). The hope is that present-day science can be refashioned along more female-friendly lines. "Ecology, as a philosophy of nature, has roots in organicism – the idea that the cosmos is an organic entity, growing and developing from within, in an integrated unity of structure and function" (100). No mention is made of religion, but one can easily see how this kind of argumentation can be molded to make mechanism anti-religion. Just no soul.

Yet, while it is certainly true that the mechanistic approach strives to present a value-free picture of the world, and it is certainly true that many mechanists want little to do with religion, organized or otherwise, this by no means implies that the mechanist has no values – or desires to see them brought to reality. Nor that mechanism, in itself, is hostile to religion. The discussions thus far in this chapter give the lie to this. The discussions of war and prejudice have been highly moral – or morally relevant. And while there are certainly going to be clashes with religion, or at least some versions of religion, much that has been said can be welcomed by the person of religion, the Christian, and used to mold and shape and correct and improve one's beliefs. Mechanism and Christian belief should not at once be categorized as unfriendly – warfare, as it is often categorized. Going the other way, organicism is often taken to be the route guided by moral concerns. It is certainly a route taken, as we have seen, by many Christians: Teilhard de Chardin, for instance. However, just as simplistic conclusions should not be drawn about mechanism, nor equally should simplistic conclusions be drawn about organicism. As the story of prejudice against the Jews shows only too clearly, organicism has been used to support some very immoral conclusions, totally antithetical to Christianity properly understood.

Prejudice against the Jews is of long standing. The Gospel of John spells out how it was the Jews who brought about the crucifixion of the Savior. "[Pilate] said to the Jews, 'Here is your King!' [15] They cried out, 'Away with him! Away with him! Crucify him!' Pilate asked them, 'Shall I crucify your King?' The chief priests answered, 'We have no king but the emperor.' [16] Then he handed him over to them to be crucified" (John

19). In the Middle Ages, for all that Jews were used as money lenders and the like, persecution was the norm. In the city of York, in the north of England, in 1190, 150 Jews holed up in Clifford's Tower and committed suicide rather than be torn apart by a howling mob. Then there was Martin Luther: "They are our public enemies. They do not stop blaspheming our Lord Christ, calling the Virgin Mary a whore, Christ, a bastard, and us changelings or abortions. If they could kill us all, they would gladly do it" (*Works* 58: 458–9). Almost paradoxically, in the nineteenth century Germany was at the forefront of allowing Jews full privileges of citizenship, such as the right to practice professions like medicine and to attend leading places of education. By the time Hitler came to power in 1933, 20% of Jews were marrying gentiles. There was nevertheless much antisemitism still. At the time Hitler lived in Vienna, the mayor Karl Lueger was notorious for his hostility to Jews. It is hardly surprising that, after the defeat in the Great War, many looked for scapegoats, and the Jews were the obvious target, with the horrendous consequences only too well known.

What was the reasoning behind this prejudice? A huge appeal of the National Socialists was their determination to give Germany that sense of ingroup, of belonging to a unity. *Ein Reich, ein Volk, ein Führer*. As the philosopher Martin Heidegger put it: "The Führer himself and only he is Reality in Germany today and in the future" (quoted in Koonz's *Nazi Conscience*, 194). That tells it all. We in the Third Reich are a living body, a whole, with each one of us a part of that vibrant organism. The *state* is an enemy. We must think rather in terms of the *nation*: a group of individuals who sacrifice personal interests for the goal of common ethnic and religious cohesion. Hitler inherited this way of thinking, and it was the foundation of everything he thought and did.

> All in all, the National Socialistic conception of state and culture is that of an organic whole. As an organic whole, the Volkish state is more than the sum of its parts, and indeed because of these parts, called individuals, are fitted together to make a higher unity, within which they in turn become capable of a higher level of life achievement, while also enjoying an enhanced sense of security. The individual is bound to this sort of freedom through the fulfillment of his duty in the service of the whole. (Quotation of party member Karl Zimmerman in 1933 in *Reenchanted Science*, 176)

A holistic ingroup. Defining itself against an outgroup. Not any outgroup. Hitler thought highly of Asians. "I have never regarded the Chinese or the Japanese as being inferior to ourselves. They belong to ancient civilizations, and I admit freely that their past history is superior to our own" (*Political Testament of Adolf Hitler*, Note #5, February 1945 to April 1945). The Jews were different. They were on the spot, and already there was a history of anti-Semitism. They were a perfect target. Hitler certainly convinced Hitler himself. After Pearl Harbor, the immediate assumption was that the main driving factor behind America's reaction was the influence of Jews on the president, Franklin Roosevelt. Five days later, Hitler declared war on the United States. The day after, December 13, 1941, Goebbels wrote in his diary: "Regarding the Jewish question, the *Führer* is determined to settle the matter once and for all. He prophesied that if the Jews once again brought about a world war, they would experience their extermination. This was not an empty phrase. The world war is here. The extermination of the Jews must be its necessary consequence" (*Jewish Enemy*, 132). The Wannsee conference – spelling out the "Final Solution" – was less than two months later.

The claim here is obviously not that organicism leads to National Socialism, which – for all the craven acceptance of many German clerics – is about as far from Christianity as it is possible to be. It is simply, as in the discussion of the implications of mechanism, a warning about drawing conclusions, too wide-ranging, too soon. The relationship between science and religion is more complex than one might have expected. Being a mechanist or being an organicist does not in itself end all discussion.

Epilogue

At the beginning of this book, I told you – warned you – that I am a historian and philosopher of science, not a theologian. You may now be feeling that you should have taken that comment a lot more seriously than you did. The cobbler should stick to his last. I have plowed blithely on without once discussing that fundamental Christian notion of substitutionary atonement – the claim that Jesus died on the Cross to forgive us our sins; and as for something like transubstantiation – the claim that during the eucharist the bread and wine turn literally into the body and blood of Christ – forget it.

I certainly will not deny that you can bring topics like these into the science–religion debate, and of course in a way I have discussed them. Supposedly, we need the sacrifice of Jesus because we are tainted with original sin – the result of Adam and Eve disobeying God and eating that wretched apple – and certainly the claim that there was no sin in the world until that event has been discussed, not very favorably to this particular assertion. Again, transubstantiation may not have been an explicit topic for debate, but the discussion of miracles is surely relevant. I very much doubt that someone who thinks a literal Resurrection is not to be endorsed is going to be a big enthusiast of water into wine, let alone wine into blood.

I am not going to apologize for the approach I have taken. If you want a more familiar approach, there is a huge amount of material out there to which you can turn. Start with the Wikipedia entry on "The Relationship between Religion and Science." Not only is it balanced and informative, but there is also a large and hugely helpful list of "Further Reading." Nor am I going to say that the approach I have taken should supersede all other approaches. My father always accused me of being too full of myself – "Michael, you were a big-head as a boy, and you are a big-head now" – but even I am not that far gone beyond modesty. This said, there is something to taking a fresh approach, a new look at things, and (here my father is proven right) I think the approach I have taken has much to commend it.

That said, why should someone interested in the science–religion relationship bother to read this book? I will give you two reasons – there may be more. First, a historian and philosopher of science is going to take science very seriously. After all, this is what we spend our lives looking at and discussing and trying to understand. I am not saying that we are going to accept uncritically everything scientists claim, but the presumption is that way. If someone comes up with a new hypothesis in science – for instance about that structure of a particular molecule ("buckyballs" come to mind) – you might at first question it, and look for evidence, and so forth, but generally you think that such claims, perhaps modified, are extending the range of our knowledge.

With respect to the topic of this book, such an attitude is going to insist that you follow in the footsteps of Thomas Henry Huxley. It is all important to start with the fact that we are modified monkeys, not modified dirt. This is not the opinion with many, especially analytic philosophers. The saintly Wittgenstein said biology had little to contribute to philosophy and went as far as to dismiss Darwin's theory in its entirety. "I have always thought that Darwin was wrong: his theory doesn't account for all this variety of species. It hasn't the necessary multiplicity." As always, gnomic remarks like this come with absolutely no hint that the speaker has ever glanced at the writings of professional evolutionists. Not that the philosophers are worse than those who come to the science–religion debate from the side of theology. The eminent theologian, the Jesuit Karl Rahner, for instance, writes a whole essay on Christianity and evolution, telling us that "the history of Nature and spirit forms an inner, graded unity in which natural history develops towards man, continues in him as his history, is conserved and surpassed in him and hence reaches its proper goal with and in the history of the human spirit." Needless to say, his reading of the professional literature is about as wide and penetrating as that of Wittgenstein.

I am not asking that the reader accept uncritically the position taken in this book about the relevance of evolution to the discussion. I am suggesting that it is a position that must be taken seriously. The second point I would make about the virtue of such an approach as mine is in important ways connected to the first point. The relevance of Darwinian evolutionary theory is hotly contested. Some would argue that it is false. Others would argue that even if it be true, it is not relevant to the science–religion debate. I – and I am sure I am not alone – have often been puzzled as to why intelligent knowledgeable people can differ so much on these things; especially why, given that today there is such overwhelming evidence in favor of a Darwinian approach to evolution, so many of those who think evolution relevant to the science–religion debate adamantly refuse to take a Darwinian stance. Teilhard de Chardin for instance: why, if he was a brilliant thinker (paleontologist), should he have so firmly opted to base his thinking on Bergson rather than Darwin?

The answer, we see now, lies in alternative root metaphors or paradigms. The Darwinians are firmly in the mechanistic tradition. Many, however, go the other way. Like Teilhard they are just as firmly within the organismic tradition. They get a directed progress upwards, they get lots of support for subsidiary claims like free will and sin, and above all they get humans triumphant – the apotheosis of the evolutionary movement. Made in the image of God. Once you see what is going on, it is all so obvious. Debates that tear us apart on the Christianity/evolution front lines are only secondarily about Christianity and evolution, Darwinism even. They are about different ways of looking at things, different ways of making sense of appearances, ways that existed centuries before either Christianity or Darwinism appeared on the scene. That makes you think. At least, it should make you think. Are we comfortable being forced into arguing in frameworks developed entirely without regard to our specific interests?

Writing books is always an exploration. You start out with an idea of what you want to say, but so very often end with something quite unexpected. That is what makes it all so very worthwhile. When I set out, I knew that the two root metaphors were important, but I had no idea how deeply they structure the whole science–religion discussion. I come at the discussion from a middle position. I am not a Christian or adherent to any other religion. Nor equally am I atheist, certainly not a New Atheist! If you are raised a Quaker, such as I, you cannot hate Christianity with the passion of a Richard Dawkins. I suppose I should be described as an "agnostic," but that doesn't really cover things for me, as, in this day and age, I don't think it really covers someone like Thomas Henry Huxley. Neither of us is indifferent to the questions, as are so many, like my wife. I have that mystical attitude of my Quaker childhood – queerer than we think it is, queerer than we could think it is. This makes me open to insights on both sides of the mechanism/organicism divide. As a mechanist, I could never reject Darwinism in the way that the organicists so often so blithely do. We have got to wrestle with the non-directionality of evolution, not ignore it or pretend it does not exist. Going the other way, even though I am not an organicist, I am deeply appreciative of process theology. My God would never be the God of Aquinas, unable to be part of human grief, suffering on the Cross in a disinterested way for sins committed long ago by others. My God would be in Bergen-Belsen, lying and suffering with Anne Frank as she died from typhoid – "delirious, terrible, burning up." He would also be the God who conquered, who made available her diary, which has been and continues to be such a huge inspiration to her fellow human beings – especially young people. "In spite of everything I still believe that people are really good at heart. I simply can't build up my hopes on a foundation consisting of confusion, misery, and death."

What a privilege it has been to write this book and share my ideas with you.

Summary of Common Misunderstandings

The life sciences and Christianity relationship is of interest/ importance only to life scientists

From the first, the life sciences – most particularly, the underlying Darwinian theory of evolution through natural selection – have had a fraught relationship with the Christian religion, starting with the fact that they deny absolutely a six-day creation, a historical Adam and Eve, and the claim that sin came into the world only because of the bad behavior of this original pair. Given the seminal importance of Darwinian thinking, not the least in understanding human nature, biological and cultural, easing this relationship is in the interest of us all.

The life sciences and Christianity relationship is of interest/ importance only to Christians

Conversely, no Christian today can afford to be ignorant of, or contemptuous towards, the life sciences. We live in an age of science and technology, where the Darwinian theory of natural selection is the dominant paradigm, the underpinning of all thinking about organisms, extinct and extant. Prima facie, there are clashes between Christianity and Darwinism. These must be worked out, not just for the comfort of Christians, but, given that here in the West we live in a predominantly Christian culture, for us all, believers and non-believers.

The life sciences and Christianity are at war

While it is certainly the case that the life sciences and Christianity have clashed, notoriously in the 1860 debate between Samuel Wilberforce, Anglican Bishop of Oxford, and morphologist Thomas Henry Huxley, Darwin's "bulldog," more commonly Christians and life scientists have found that their differences are soluble, and indeed can reflect favorably on both religion and science.

Christianity is just about ethics

Much of the science–religion trouble stems from the fact that Christianity is not just about ethics. It makes factual claims, for instance that Adam and Eve were uniquely the original pair and free of sin. Unless factual claims like these can be reinterpreted (metaphorically), Christianity will continue to clash with the life sciences.

The life sciences are just about facts

The life sciences, evolution in particular, have relevant implications for human nature, explaining our thinking and behavior in terms of our past. Moral and social policies about humans, for instance about sexual behavior, depend intimately on understanding such implications.

The life sciences have no debts to Christianity

Much in Darwin's theory started with claims by Christians, for instance that organisms are not random but adapted, "as if" designed, that there is a tree of life sketched out by our evolutionary past, and – more controversially (in the opinion of Richard Dawkins but not Charles Darwin) – that humans have a special status.

Christianity has no debts to the life sciences

Sophisticated Christianity today interprets many of the claims of the Bible metaphorically in the light of evolutionary theory. These include a time span of billions of years not six days, a gradual development of a group of humans not miraculous creation of but two, and a realization that sin (as well as love) is part of human nature and not an immediate consequence of the bad behavior of the original pair.

Christianity is for people who are scared of reality

Many of the leading evolutionists, prepared to go where their science leads them, have been committed Christians. These include the populational geneticist Ronald A. Fisher, the Russian-born fruit-fly specialist Theodosius Dobzhansky, and today the eminent paleontologist Simon Conway Morris.

The life sciences are for people who are tough-minded realists

Some have turned to the life sciences precisely because they seek a meaning to life beyond just crude uninterpreted factual discoveries. Starting with Darwin and Huxley, many have been agnostics, amazed and inspired by our understanding of the living world, and somewhat mystical as to whether there is any ultimate meaning to human existence.

This book is the final word on the topic

Prove me wrong. Write your own!

References and Further Reading

Biblical quotations are from the Authorized King James Version.

Chapter 1

Aquinas, St T. 1952. *Summa Theologica, I*. London: Burns, Oates and Washbourne.

Augustine [396] 1998. *Confessions*. (Translator H. Chadwick) Oxford: Oxford University Press.

Barnes, J. (editor) 1984. *The Complete Works of Aristotle*. Princeton: Princeton University Press.

Boyle, R. [1686] 1996. *A Free Enquiry into the Vulgarly Received Notion of Nature*. (Editors E. B. Davis and M. Hunter) Cambridge: Cambridge University Press.

Boyle, R. 1688. *A Disquisition about the Final Causes of Natural Things: Wherein it is Inquir'd Whether, And (if at all) With what Cautions, a Naturalist should admit Them?* London: John Taylor.

Calvin, J. 1536. *Institutes of the Christian Religion*. Grand Rapids, MI: Eerdmans.

Cooper, J. M. (editor) 1997. *Plato: Complete Works*. Indianapolis: Hackett.

Darwin, C. 1859. *On the Origin of Species by Means of Natural Selection, or the Preservation of Favoured Races in the Struggle for Life*. London: John Murray.

Darwin, C. 1987. *Charles Darwin's Notebooks, 1836–1844*. (Editors P. H. Barrett, P. J. Gautrey, S. Herbert, D. Kohn, and S. Smith) Ithaca, NY: Cornell University Press.

Darwin, E. [1794–1796] 1801. *Zoonomia; or, The Laws of Organic Life*. 3rd ed. London: J. Johnson.

Darwin, E. 1803. *The Temple of Nature*. London: J. Johnson.

de Buffrénil, V., J. O. Farlow, and de Ricqlès, A. 1986. Growth and function of stegosaurus plates: evidence from bone histology. *Paleobiology* 12: 459–73.

Dijksterhuis, E. J. 1961. *The Mechanization of the World Picture*. Oxford: Oxford University Press.

Dobzhansky, T. 1937. *Genetics and the Origin of Species*. New York: Columbia University Press.

Dupré, J. 2003. *Darwin's Legacy: What Evolution Means Today*. Oxford: Oxford University Press.

Evangelical Lutheran Church in America. 1996. Basis for our caring. *This Sacred Earth: Religion, Nature, Environment*. (Editor R. S. Gottlieb) New York: Routledge.

Farlow, J. O., C. V. Thompson, and D. E. Rosner. 1976. Plates of the dinosaur Stegosaurus: forced convection heat loss fins? *Science* 192: 1123–25.

Halstead, B. 1975. *The Evolution and Ecology of the Dinosaurs*. London: Peter Lowe.

Huxley, J. 1942. *Evolution: The Modern Synthesis*. London: Allen and Unwin.

Kant, I. [1790] 2000. *Critique of the Power of Judgment*. (Editor P. Guyer) Cambridge: Cambridge University Press.

Kuhn, T. 1962. *The Structure of Scientific Revolutions*. Chicago: University of Chicago Press.

Lamarck, J. B. 1809. *Philosophie zoologique*. Paris: Dentu.

Nagel, T. 2012. *Mind and Cosmos: Why the Materialist Neo-Darwinian Conception of Nature Is Almost Certainly False*. New York: Oxford University Press.

New Catholic Encyclopedia. 2002. 2nd ed. Farmington Hills, MI: Gale Research Inc.

Paley, W. [1802] 1819. *Natural Theology (Collected Works: IV)*. London: Rivington.

Ruse, M. 1996. *Monad to Man: The Concept of Progress in Evolutionary Biology*. Cambridge, MA: Harvard University Press.

Schelling, F. W. J. 1833–34 [2008]. *On the History of Modern Philosophy*. Cambridge: Cambridge University Press.

Schelling, F. W. J. 1988. *Ideas for a Philosophy of Nature – as Introduction to the Study of this Science 1797 – second edition 1803*. (Translators E. E. Harris and P. Heath) Cambridge: Cambridge University Press.

Sedley, D. 2008. *Creationism and Its Critics in Antiquity*. Berkeley, CA: University of California Press.

Smuts, J. C. 1926. *Holism and Evolution*. London: Macmillan.

Spencer, H. 1860. The social organism. *Westminster Review* LXXIII: 90–121.

Spencer, H. [1857] 1868. Progress: Its law and cause. *Westminster Review* LXVII: 244–67.

Spencer, H. 1879. *The Data of Ethics*. London: Williams and Norgate.

Watson, J. D., and F. H. C. Crick. 1953. Molecular structure of nucleic acids. *Nature* 171: 737–8.

Whewell, W. 1837. *The History of the Inductive Sciences* (3 vols). London: Parker.

Whitehead, A. N. 1926. *Science and the Modern World*. Cambridge: Cambridge University Press.

Wright, S. 1931. Evolution in Mendelian populations. *Genetics* 16: 97–159.

Wright, S. 1932. The roles of mutation, inbreeding, crossbreeding and selection in evolution. *Proceedings of the Sixth International Congress of Genetics* 1: 356–66.

Chapter 2

Anon. [Ansted, D. T.] 1860a. Natural selection. *All the Year Round* 3: 293–9.

Anon. [Ansted, D. T.] 1860b. Species. *All the Year Round* 3: 174–8.

Anselm, St. 1903. *Proslogium, Monologium, An Appendix in Behalf of the Fool by Gaunilon; and Cur Deus Homo*. (Translator S. N. Deane). Chicago: Open Court.

Aquinas, St T. 1952. *Summa Theologica, I*. London: Burns, Oates and Washbourne.

Baynes, T. S. 1873. Darwin on expression. *Edinburgh Review* 137: 492–508.

Behe, M. 1996. *Darwin's Black Box: The Biochemical Challenge to Evolution*. New York: Free Press.

Biological Sciences Curriculum Study. 1963. *Biological Science: Molecules to Man*. Boston: Houghton Mifflin.

Calvin, J. 1536. *Institutes of the Christian Religion*. Grand Rapids, MI: Eerdmans.

Collins, R. 1999. A scientific argument for the existence of God: The fine-tuning argument. *Reason for the Hope Within*. (Editor M. Murray) 47–75. Grand Rapids, MI: Eerdmans.

Coyne, J. A. 2015. *Faith Versus Fact: Why Science and Religion Are Incompatible*. New York: Viking.

Darwin, C. 1868. *The Variation of Animals and Plants Under Domestication*. London: Murray.

Darwin, C. 1958. *The Autobiography of Charles Darwin, 1809–1882*. (Edited by Nora Barlow) London: Collins.

Dawkins, R. 1983. *Universal Darwinism. Evolution from Molecules to Men*. (Editor D. S. Bendall) 403–25. Cambridge: Cambridge University Press.

Dawkins, R. 1986. *The Blind Watchmaker*. New York, NY: Norton.

Descartes, R. 1964. *Meditations. Philosophical Essays*, 59–143. Indianapolis: Bobbs-Merrill.

Draper, J. W. 1875. *History of the Conflict Between Religion and Science*. New York: Appleton.

Haldane, J. B. S. 1932. *The Causes of Evolution*. New York: Cornell University Press.

Heidegger, M. 1959. *An Introduction to Metaphysics*. New Haven: Yale University Press.

Hick, J. 1973. *God and the Universe of Faiths: Essays in the Philosophy of Religion*. New York: St Martin's Press.

Hume, D. [1779] 1990. *Dialogues Concerning Natural Religion*. (Editor M. Bell) London: Penguin.

Huxley, J. 1912. *The Individual in the Animal Kingdom*. Cambridge: Cambridge University Press.

Huxley, T. H. 1893. *Evolution and Ethics with a New Introduction*. (Edited by M. Ruse) Princeton: Princeton University Press.

Jacob, F. 1977. Evolution and tinkering. *Science* 196: 1161–66.

Keats, J. 1819. Letter to siblings 1819. *The Letters of John Keats, 1814–1821*. (Edited by Hyder Edward Rollins), 100–4. Cambridge, MA: Harvard University Press.

Kierkegaard, S. 1992. *Concluding Unscientific Postscript to Philosophical Fragments*, Vol. 1 (Kierkegaard's Writings, Vol. 12.1) (Translators H. V. Hong, and E. H. Hong) Princeton, NJ: Princeton University Press.

Lewis, C. S. 1942. *The Screwtape Letters*. London: Geoffrey Bles.

Lewis, C. S. 1950–1956. *The Chronicles of Narnia Set*. London: Geoffrey Bles.

Lewis, C. S. 1955. *Surprised by Joy: The Shape of My Early Life*. London: Geoffrey Bles.

Meléndez-Hevia, E., T. G. Waddell, and M. Cascante. 1996. The puzzle of the Krebs citric acid cycle: assembling the pieces of chemically feasible reactions, and opportunism in the design of metabolic pathways during evolution. *Journal of Molecular Evolution* 43: 293–303.

Miller, K. 1999. *Finding Darwin's God*. New York: Harper and Row.

Moore, A. 1890. The Christian doctrine of God. *Lux Mundi*. (Editor C. Gore) 57–109. London: John Murray.

Nagel, T. 2012. *Mind and Cosmos: Why the Materialist Neo-Darwinian Conception of Nature Is Almost Certainly False*. New York: Oxford University Press.

Newman, J. H. 1973. *The Letters and Diaries of John Henry Newman*, XXV. (Editors C. S. Dessain and T. Gornall) Oxford: Clarendon Press.

Plantinga, A. 1993. *Warrant and Proper Function*. Oxford: Oxford University Press.

Weinberg, S. 1999. A designer universe? *New York Review of Books* 46: 46–8.

Whewell, W. 1837. *The History of the Inductive Sciences* (3 vols). London: Parker.

Whitcomb, J. C. and H. M. Morris. 1961. *The Genesis Flood: The Biblical Record and Its Scientific Implications*. Philadelphia: Presbyterian and Reformed Publishing Company.

White, A. D. 1896. *History of the Warfare of Science with Theology in Christendom*. New York: Appleton.

Chapter 3

Aquinas, St T. 1952. *Summa Theologica, I*. London: Burns, Oates and Washbourne.

Arthur, W. 2021. *Understanding Evo-Devo*. Cambridge: Cambridge University Press.

Bergson, H. 1907. *L'évolution créatrice*. Paris: Alcan.

Bergson, H. 1911. *Creative Evolution*. New York: Holt.

Berry, T. 2009. *The Sacred Universe: Earth, Spirituality, and Religion in the Twenty-First Century*. (Editor E.-M. Tucker) New York: Columbia University Press.

Clifford, W. K. 1901. Body and mind (from *Fortnightly Review*). *Lectures and Essays of the Late William Kingdom Clifford* Vol. 2. (Editors L. Stephen and F. Pollock) 1–51. London: Macmillan.

Fisher, R. A. 1930. *The Genetical Theory of Natural Selection*. Oxford: Oxford University Press.

Fodor, J. and M. Piattelli-Palmarini. 2010. *What Darwin Got Wrong*. New York: Farrar, Straus, and Giroux.

Gayon, J. 2013. Darwin and Darwinism in France after 1900. *The Cambridge Encyclopedia of Darwin and Evolutionary Thought*. (Editor M. Ruse) 300–312. Cambridge: Cambridge University Press.

Greene, J. C. and M. Ruse. 1996. On the nature of the evolutionary process: the correspondence between Theodosius Dobzhansky and John C. Greene. *Biology and Philosophy* 11: 445–91.

Gunn, J. A. 1920. *Bergson and His Philosophy*. New York: E. P. Dutton.

Laland, K., J. Endler, and J. Odling-Smee. 2017. Niche construction, sources of selection and trait coevolution. *Interface Focus (Royal Society)* https://royalso cietypublishing.org/doi/10.1098/rsfs.2016.0147

Laland, K., T. Uller, M. Feldman et al. 2014. Does evolutionary theory need a rethink? *Nature* 514: 161–4.

Lewontin, R. C. 1974. *The Genetic Basis of Evolutionary Change*. New York, NY: Columbia University Press.

Spencer, H. 1862. *First Principles*. London: Williams and Norgate.

Leibniz, G. F. W. 1714. *Monadology and other Philosophical Essays*. New York: Bobbs-Merrill.

Medawar, P. B. 1961. Review of *The Phenomenon of Man*. *Mind* 70: 99–106.

O'Connell, G. 2017. Will Pope Francis remove the Vatican's "warning" from Teilhard de Chardin's writings? *America* November 21.

O'Connell, J., and M. Ruse. 2019. After Darwin: Morality in a secular world. *Secular Studies* 1: 161–85.

Prince of Wales, T. Juniper, and I. Skelly. 2010. *Harmony: A New Way of Looking at Our World*. New York: Harper Collins.

Rogers, P. 2001. *Song of the World Becoming: New and Collected Poems 1981– 2001*. Minneapolis: Milkweed.

Schelling, F. W. J. 1988. *Ideas for a Philosophy of Nature – as Introduction to the Study of this Science 1797 – second edition 1803*. (Translators E. E. Harris and P. Heath) Cambridge: Cambridge University Press.

Teilhard de Chardin, P. 1955. *The Phenomenon of Man*. London: Collins.

Whitehead, A. N. 1929 [1978]. *Process and Reality: An Essay in Cosmology*. New York: Free Press.

Chapter 4

Aquinas, St T. 1952. *Summa Theologica, I*. London: Burns, Oates and Washbourne.

Bergson, H. 1911. *Creative Evolution*. New York: Holt.

Carnegie, A. 1889. The Gospel of Wealth. *North American Review* 148: 653–65.

Cooper, J. M. (editor). 1997. *Plato: Complete Works*. Indianapolis, IN: Hackett.

Darwin, C. 1859. *On the Origin of Species by Means of Natural Selection, or the Preservation of Favoured Races in the Struggle for Life*. London: John Murray.

Darwin, C. 1871. *The Descent of Man*. London: John Murray.

Dennett, D. C. 1984. *Elbow Room: The Varieties of Free Will Worth Wanting*. Cambridge, MA: MIT Press.

Dickens, C. [1853] 1948. *Bleak House*. Oxford: Oxford University Press.

Edwards, J. 1969. *Freedom of the Will*. (Editor A. S. Kaufman) Indianapolis, IN: Bobbs-Merrill Co.

Fischer, J. M., R. Kane, D. Pereboom, and M. Vargas. 2007. *Four Views on Free Will*. Maldan, MA: Blackwell.

Hume, D. [1739–1740] 1978. *A Treatise of Human Nature*. Oxford: Oxford University Press.

Hume, D. [1751] 1983. *An Enquiry Concerning the Principles of Morals*. Indianapolis: Hackett.

Huxley, J. S. 1948. *UNESCO: Its Purpose and Its Philosophy*. Washington, DC: Public Affairs Press.

Huxley, T. H. 1893. *Evolution and Ethics with a New Introduction*. (Edited by M. Ruse) Princeton: Princeton University Press.

Huxley, T. H. and J. Huxley. 1947. Evolutionary ethics. *Touchstone for Ethics*. New York: Harper.

Kant, I. 1781 [1899]. *Critique of Pure Reason*. (Translator J. M. D. Meikeljohn) New York: Colonial Press.

Mackie, J. L. 1977. *Ethics*. Harmondsworth: Penguin.

Maine, H. J. S. 1861. *Ancient Law; Its Connection to the Early History of Society, and Its Relation to Modern Ideas*. London: John Murray.

Moore, G. E. 1903. *Principia Ethica*. Cambridge: Cambridge University Press.

Sidgwick, H. 1876. The theory of evolution in its application to practice. *Mind* 1: 52–67.

Singer, P. 1972. Famine, affluence and morality. *Philosophy and Public Affairs* 1: 229–43.

Spencer, H. 1851. *Social Statics: Or, the Conditions Essential to Human Happiness Specified, and the First of Them Developed*. London: Chapman.

Spencer, H. [1857] 1868. Progress: Its law and cause. *Westminster Review* LXVII: 244–67.

Wilson, D. S. and E. O. Wilson 2007. Rethinking the theoretical foundation of sociobiology. *Quarterly Review of Biology* 82: 327–48.

Wilson, E. O. 1975. *Sociobiology: The New Synthesis*. Cambridge, MA: Harvard University Press.

Chapter 5

Agassiz, E. C. (editor). 1885. *Louis Agassiz: His Life and Correspondence*. Boston: Houghton Mifflin.

Agricola, G. [1556] 1950. *De Re Metallica*. (Translators H. C. Hoover and L. C. Hoover) New York: Dover.

Capra, F. 1987. Deep Ecology: a new paradigm. *Deep Ecology for the 21st Century*. (Editor G. Sessions) Boston, MA: Shambhala.

Darwin, C. R. 1839. Observations on the parallel roads of Glen Roy, and other parts of Lochaber in Scotland, with an attempt to prove that they are of marine origin. *Philosophical Transactions of the Royal Society of London* https://doi.org/10.1098/rstl.1839.0005

Dickinson, E. 1960. *The Complete Poems of Emily Dickinson*. New York: Little, Brown.

Francis (Pope). 2015. *Encyclical on Climate Change and Inequality: On Care for Our Common Home*. New York: Melville House.

Hutton, J. 1788. Theory of the Earth; or an Investigation of the Laws observable in the Composition, Dissolution, and Restoration of Land upon the Globe. *Transactions of the Royal Society of Edinburgh* 1: 209–304.

International Rice Research Institute. 2007. *The Great Rice Robbery*. Penang, Malaysia: Pesticide Action Network.

Kant, I. [1785] 2007. *Groundwork for the Metaphysic of Morals*. www.earlymoderntexts.com/assets/pdfs/kant1785.pdf

Lovelock, J. E. 1979. *Gaia: A New Look at Life on Earth*. Oxford: Oxford University Press.

Lovelock, J. E., and L. Margulis. 1974a. Homeostatic tendencies of the Earth's atmosphere. *Origins of Life* 5: 93–103.

Lovelock, J. E. and L. Margulis. 1974b. Atmospheric homeostasis by and for the biosphere: the Gaia Hypothesis. *Tellus* 26: 1–10.

Miles, M., and V. Shiva. 2014. *Ecofeminism (Critique. Influence. Change)*. London: Zed Books.

Moncrief, L. 1970. The cultural basis of our environmental crisis. *Science* 170: 508–12.

Naydler, J. (editor). 1996. *Goethe on Science: An Anthology of Goethe's Scientific Writings*. Edinburgh: Floris Books.

Razak, A. 1990. Toward a womanist analysis of birth. *Reweaving the World: The Emergence of Ecofeminism*. (Editors I. Diamond and G. F. Orenstein) 165–72. San Francisco: Sierra Club.

Regis, E. 2019. *Golden Rice: The Imperiled Birth of a GMO Superfood*. Baltimore: Johns Hopkins University Press.

Rolston, H. III, 1988. *Environmental Ethics*. Philadelphia: Temple University Press.

Smuts, J. C. 1926. *Holism and Evolution*. London: Macmillan.

Spretnak, C. 1989. Toward an ecofeminist spirituality. *Healing the Wounds: The Promise of Ecofeminism*. (Editor J. Plant) 127–32. Philadelphia: New Society Publishers.

Taylor, P. W. 1981. The ethics of respect for nature. *Environmental Ethics* 3: 197–218.

Van Dyke, F., D. C. Mahan, J. K. Sheldon and R. H. Brand. 1996. *Redeeming Creation: The Biblical Basis for Environmental Stewardship*. Downers Grove, IL: InterVarsity Press.

White, L. 1967. The historical roots of our ecological crisis. *Science* 155: 1203–7.

Whitehead, A. N. 1926. *Science and the Modern World*. Cambridge: Cambridge University Press.

Wilson, E. O. 1984. *Biophilia*. Cambridge, MA: Harvard University Press.

Zell-Ravenheart, O. 2009. *Green Egg Omelet: An Anthology of Art and Articles from the Legendary Pagan Journal*. Franklin Lakes, NJ: New Page Books.

Chapter 6

Allport, G. [1954] 1958. *The Nature of Prejudice*. Garden City, NY: Doubleday.

Ardrey, R. 1961. *African Genesis: A Personal Investigation into the Animal Origins and Nature of Man*. New York: Atheneum.

Bang, J. P. 1917. *Hurrah and Hallelujah: The Teaching of Germany's Poets, Prophets, Professors and Preachers*. New York: George H. Doran.

Baxter, R. 2021. *A Christian Directory, Or, a Body of Practical Divinity and Cases of Conscience: Christian Ethics (Or Private Duties)*. Los Angeles: HardPress Publishing.

Bennhold, K. and M. Pronczuk. 2022. Poland shows the risks for women when abortion is banned. *New York Times* www.nytimes.com/2022/06/12/world/eu rope/poland-abortion-ban.html

Brewer, M. B. 1999. The Psychology of Prejudice: Ingroup Love or Outgroup Hate? *Journal of Social Issues* 55: 429–44.

Ceci, S. J. and W. M. Williams. 2009. *The Mathematics of Sex: How Biology and Society Conspire to Limit Talented Women and Girls*. New York: Oxford University Press.

Chagnon, N. 1988. Life histories, blood revenge, and warfare in a tribal population. *Science* 239: 985–92.

Cicero, M. T. 1913. *De Officiis*. (Translator W. Miller) Cambridge, MA: Harvard University Press.

Clark, C. 2009. *Iron Kingdom: The Rise and Downfall of Prussia, 1600–1947*. Cambridge, MA: Harvard University Press.

Crook, P. 1994. *Darwinism, War and History: The Debate over the Biology of War from the "Origin of Species" to the First World War*. Cambridge: Cambridge University Press.

Dickens, C. 1857. *Little Dorrit*. London: Bradbury and Evans.

Dart, R. 1953. The predatory transition from ape to man. *International Anthropological and Linguistic Review* 1: 201–17.

Darwin, C. 1871. *The Descent of Man*. London: John Murray.

Edsall, T. B. November 1, 2018. The Trump legions. *New York Times*, www.nytimes.com/2018-11/01/opinion/the-trump-legions.html.

Ferguson, R. B. 2013. The prehistory of war and peace in Europe and the Near East. *War, Peace, and Human Nature: The Convergence of Evolutionary and Cultural Views*. (Editor D. P. Fry) 191–240. Oxford: Oxford University Press.

Ferguson, R. B. 2015. History, explanation, and war among the Yanomami: a response to Chagnon's Noble Savages. *Anthropological Theory* 15: 377–406.

Fichte, J. G. 1821 [1922]. *Addresses to the German Nation*. Chicago: Open Court.

Grossman, D. 2009. *On Killing: The Psychological Cost of Learning to Kill in War and Society*. New York: Back Bay Books.

Harrington, A. 1996. *Reenchanted Science: Holism in German Culture from Wilhelm II to Hitler*. Princeton, NJ: Princeton University Press.

Hegel, G. W. F. 1821 [1991]. *Elements of the Philosophy of Right*. (Editor A. Wood) Cambridge: University of Cambridge Press.

Heidegger, M. 1959. *An Introduction to Metaphysics*. New Haven: Yale University Press.

Herf, J. 2006. *The Jewish Enemy: Nazi Propaganda during World War II and the Holocaust*. Cambridge, MA: Harvard University Press.

Hersch, S. M. 1972. *Cover-up: The Army's Secret Investigation of the Massacre at My Lai 4*. New York: Random House.

Hitler, A. [1925] 2009. *Mein Kampf* (Translator M. Ford) www.hitler-library.org/

Holmes, A. F. (editor). 2005. *War and Christian Ethics*. 2nd ed. Grand Rapids, MI: Baker.

Jenkins, P. 2014. *The Great and Holy War: How World War I Became a Religious Crusade*. New York: HarperOne.

Jones, E. 1849. *The Land Monopoly: The Suffering and Demoralization Caused by It, and the Justice & Expediency of Its Abolition*. London: Charles Fox.

Koonz, C. 2003. *The Nazi Conscience*. Cambridge, MA: Belknap.

Luther, M. 1915. *Works*. Philadelphia, PA: Holman.

Merchant, C. 1980. *The Death of Nature: Women, Ecology, and the Scientific Revolution: A Feminist Reappraisal of the Scientific Revolution*. Scranton, PA: HarperCollins.

Pinker, S. 2011. *The Better Angels of Our Nature: Why Violence Has Declined*. New York: Viking.

Reich, D. 2018. *Who We Are and How We Got Here: Ancient DNA and the New Science of the Human Race*. New York: Pantheon.

SBCEC (Southern Baptist Convention Executive Committee). 2022. Southern Baptists and Women Pastors. *Baptist 2 Baptist*: www.baptist2baptist.net/print friendly.asp?ID=58

Scott, E. 2019. Trump's most insulting – and violent – language is often reserved for immigrants. *Washington Post*: www.washingtonpost.com/polit ics/2019-10/02/trumps-most-insulting-violent-language-is-often-reserved-immigrants/

Selous, E. 1901. An observational diary of the habits – mostly domestic – of the great crested grebe (*Podicipes cristatus*). Continued as: An observational diary of the habits – mostly domestic – of the great crested grebe (*Podicipes cristatus*), and of the peewit (*Vanellus vulgaris*), with some general remarks. *Zoologist*: 5: 161–83; 5: 339–50; 5: 454–62; 6: 133–44.

Smith H. M., D. Chiszar, and R. R. Montanucci. 1997. Subspecies and classification. *Herpetological Review* 28: 13–16.

Von Bernhardi, F. 1912. *Germany and the Next War*. London: Edward Arnold.

Von Bernhardi, F. 1914. *Britain as Germany's Vassal*. London: Dawson.

Von Bernhardi, F. 1920. *Vom Kriege der Zukunft. Nach den Erfahrungen des Weltkrieges*. Berlin: Mittler.

Epilogue

Frank, A. 1952. *The Diary of a Young Girl*. New York: Doubleday and Co.

Rahner, K. 1966. Christology within an evolutionary view of the world. *Theological Investigations. V. Later Writings*, 157–92. Baltimore: Helicon Press.

Wittgenstein, L. 1980. *Culture and Value*. (Editor G. H. Von Wright) Oxford: Blackwell.

Further Reading

Aquinas, St T. [1259–1265] 1975. *Summa Contra Gentiles*. (Translator V. J. Bourke) Notre Dame: University of Notre Dame Press.

Augustine, St. [413–426] 1998. *The City of God against the Pagans*. (Editor and translator R. W. Dyson) Cambridge: Cambridge University Press.

Augustine. 1982. *The Literal Meaning of Genesis*. (Translator J. H. Taylor) New York: Newman.

Bellah, R. 2011. *Religion in Human Evolution: From the Paleolithic to the Axial Age*. Cambridge, MA: Harvard University Press.

Brooks, R. 2021. Darwin's closet: the queer sides of *The Descent of Man* (1871). *Zoological Journal of the Linnean Society* 191: 323–46.

Clark, E. D. 2006. Abraham and Sarah: a love story without end. *Meridian Magazine*: https://latterdaysaintmag.com/

Clark, R. W. 1960. *Sir Julian Huxley, F.R.S.* London: Phoenix House.

Darwin, C. 1839. *Journal of Researches into the Geology and Natural History of the Various Countries Visited by HMS Beagle.* London: Henry Colburn.

Darwin, C. 1985–. *The Correspondence of Charles Darwin.* Cambridge: Cambridge University Press.

Fiske, A. P. and T. S. Rai. 2014. *Virtuous Violence: Hurting and Killing to Create, Sustain, End, and Honor Social Relationships.* Cambridge: Cambridge University Press.

Fry, D. P. 2013. War, peace, and human nature: the challenge of achieving scientific objectivity. *War, Peace, and Human Nature: The Convergence of Evolutionary and Cultural Views.* (Editor D. P. Fry) 1–21. Oxford: Oxford University Press.

Gibson, A. 2013. Edward O. Wilson and the organicist tradition. *Journal of the History of Biology* 46: 599–630.

Gould, S. J. 1999. *Rocks of Ages: Science and Religion in the Fullness of Life.* New York: Ballantine.

Grayling, A. C. 2006. *Among the Dead Cities: Is the Targeting of Civilians in War ever Justified?* London: Bloomsbury.

Hollum, J. R. 1987. *Elements of General and Biological Chemistry.* New York: Wiley.

Huxley, J. S. 1943. *TVA: Adventure in Planning.* London: Scientific Book Club.

Johnson, P. E. 1991. *Darwin on Trial.* Washington, DC: Regnery Gateway.

Kant, I. [1785] 2007. *Groundwork for the Metaphysic of Morals.* www .earlymoderntexts.com/assets/pdfs/kant1785.pdf

Larson, E. J. 1997. *Summer for the Gods: The Scopes Trial and America's Continuing Debate over Science and Religion.* New York: Basic Books.

Lucas, J. R. 1979. Wilberforce and Huxley: A legendary encounter. *Historical Journal* 22: 313–30.

MacArthur, R. H., and E. O. Wilson. 1967. *The Theory of Island Biogeography.* Princeton, NJ: Princeton University Press.

Mathez, E. A., and J. E. Smerdon. 2018. *Climate Change: The Science of Global Warming and Our Energy Future*. New York: Columbia University Press.

Maynard Smith, J. 1964. Group selection and kin selection. *Nature* 201: 1145–7.

Morgan, S. R. 1990. Schelling and the origins of his *Naturphilosophie*. In *Romanticism and the Sciences*. (Edited by A. Cunningham and N. Jardine) 25–37. Cambridge: Cambridge University Press.

Naess, A. 1986. The deep ecological movement: Some philosophical aspects. *Deep Ecology in the 21st Century*. (Editor G. Sessions) Boston: Shambhala.

Newson, L., and P. J. Richerson. 2021. *The Story of Us: A New Look at Human Evolution*. Oxford: Oxford University Press.

Noll, M. 2002. *America's God: From Jonathan Edwards to Abraham Lincoln*. New York: Oxford University Press.

Numbers, R. L. 2006. *The Creationists: From Scientific Creationism to Intelligent Design*. Standard ed. Cambridge, MA: Harvard University Press.

O'Brien, W. V. 1992. Desert Storm: A just war analysis. *St. John's Law Review* 66: 797–823.

Pennock, R. 1998. *Tower of Babel: Scientific Evidence and the New Creationism*. Cambridge, MA: MIT Press.

Provine, W. B. 1971. *The Origins of Theoretical Population Genetics*. Chicago: University of Chicago Press.

Richards, R. J. 2003. *The Romantic Conception of Life: Science and Philosophy in the Age of Goethe*. Chicago: University of Chicago Press.

Richards, R. J., and M. Ruse. 2016. *Debating Darwin*. Chicago: University of Chicago Press.

Roberts, S., and M. Rizzo. 2020. The psychology of American racism. *OSF Preprints* https://doi.org/10.31219/osf.io/w2h73

Ruse, M. (editor). 1988. *But Is It Science? The Philosophical Question in the Creation/Evolution Controversy*. Buffalo, NY: Prometheus.

Ruse, M. 2010. *Science and Spirituality: Making Room for Faith in the Age of Science*. Cambridge: Cambridge University Press.

Ruse, M. 2021a. *A Philosopher Looks at Human Beings*. Cambridge: Cambridge University Press.

Ruse, M. 2021b. The Arkansas creationism trial forty years on. *Karl Popper's Science and Philosophy*. (Editors P. Zuzana, and D. Merritt) 257–76. Switzerland: Springer.

Ruse, M. 2022a. *Understanding Natural Selection*. Cambridge: Cambridge University Press.

Ruse, M. 2022b. *Why We Hate: Understanding the Roots of Human Conflict*. Oxford: Oxford University Press.

Ruse, M. 2023. Evolutionary ethics: The wrong road taken. *"Mind": The Early Years*. (Editor L. M. Verburgt) (forthcoming).

Sedley, D. 2008. *Creationism and Its Critics in Antiquity*. Berkeley: University of California Press.

Sobolewska, M. and R. Ford. 2020. *Brexitland*. Cambridge: Cambridge University Press.

Stanley, J. 2018. *How Fascism Works: The Politics of Us and Them*. New York: Random House.

Trivers, R. L. 1971. The evolution of reciprocal altruism. *Quarterly Review of Biology* 46: 35–57.

Tuttle, R. H. 2014. *Apes and Human Evolution*. Cambridge, MA: Harvard University Press.

Whistler, D. and J. Kahl. 2014. Two poems by F. W. J. Schelling. *Clio: A Journal of Literature, History, and the Philosophy of History* 43: 177–96.

Figure Credits

Index

Printed in the United States
by Baker & Taylor Publisher Services